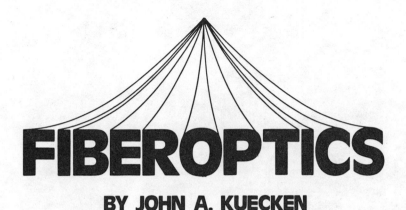

FIBEROPTICS

BY JOHN A. KUECKEN

TAB BOOKS Inc.

BLUE RIDGE SUMMIT, PA. 17214

FIRST EDITION

THIRD PRINTING

Printed in the United States of America

Library of Congress Cataloging in Publication Data

Kuecken, John A.
    Fiberoptics.

    Includes index.
    1. Fiber optics. I. Title.
TA1800.K83   621.36'92   80-19880
ISBN 0-8306-9709-8
ISBN 0-8306-1236-X (pbk.)

Introduction

Ever since May 24, 1844, when Samuel F.B. Morse sent the historic message "What hath God wrought" from the Supreme Court chambers in Washington, DC and had the message echoed back by Alfred Vail located in Baltimore, MD, men have sought to make the transfer of information between distant places faster, cheaper, and more convenient. Morse had solved the problem of conveying information over great distances in the form of electrical impulses traveling over a single wire at nearly the speed of light, with a few delays added by repeaters and sounders.

Within a few years, "lightning slingers" had developed their skills to the point where plain text messages could be sent and copied at 40 to 50 words per minute. On October 24, 1861, Western Union work crews from California and Nebraska met in Salt Lake City, UT, and the first transcontinental telegram was sent by Stephen J. Field, chief justice of California, to President Abraham Lincoln, declaring California's loyalty to the Union. In 1874, Thomas Alva Edison developed the quadruplex system which permitted the transmission of four different messages simultaneously over the same telegraph wire.

In many respects, the rapidly growing science or art of fiberoptic devices is a direct lineal descendent of the telegraph. Dr. Charles K. Kao, a pioneer in the field of fiberoptics, is quoted as saying, "The stage is set for an enrichment of life like that following the invention of the steam engine, the light bulb, and the transistor." And Dr. Kao is not alone in his assessment. Industrial giants such as

Hewlett Packard, Corning Glass, Motorola, and others are betting heavily upon the future of fiberoptics. The Bell System and ITT World Communications are using fiberoptic cables in demonstration experiments to replace telephone trunk cables. In short, the stage is being set for yet another technical revolution.

This text is aimed at providing a practical working knowledge of fiberoptic devices for engineers and technicians who may choose to ride this wave of the future. It is intended to be primarily practical and to this end, the mathematical discourses have been constrained to those areas of practical significance to the would-be practitioner.

There is little doubt in my mind that many of the dreams of the proponents of fiberoptics will come to pass. The main question is the rate at which this growth will progress. Will some political force accelerate the growth, as the Civil War accelerated the transcontinental telegraph? Or instead will a chronic energy shortage and sagging economy under spiraling inflation be unable to afford the wonders in store? Time alone will answer these questions. However, for the prudent man, forewarned is forearmed. I hope that you will find the prospects discussed in the text as interesting as I did while doing the research for this book.

John A. Kuecken

Contents

Other TAB books by the author:

1

Why Fiberoptics?

Even though the art of fiberoptic devices is still in its infancy, the antecedents stretch back a long time. If you have ever seen someone doing lacy Venetian glasswork, you will recall seeing the light from the incandescent end of the filament propagating and bending through the filament to glow like a tiny star at the other end. Because the art of Venetian glassblowing dates to the Renaissance, it seems likely that a great many people actually witnessed fiberoptic propagation long before anyone made any practical use of the phenomenon.

The first generally accepted demonstration of the fact that light will follow a curved transparent medium is credited to John Tyndall. In 1870, at a lecture before the British Royal Society, Tyndall showed this phenomenon. His apparatus consisted of a glass beaker with a spout on the bottom. The beaker was filled with water and strongly illuminated from the side opposite the spout. When the cork was removed from the spout, the water arced down into the second container which was shielded from the strong light. The light could be seen to arch down into the second container which was then illuminated by the light bending with the stream of water.

This was a bit of a pill to swallow; after all, the fact that light travels in straight lines was well known and established. It could bounce off something and change direction by being reflected. It could also bend sharply when going from one medium into another as from air into water or air into glass. This was a widely known and mathematically describable phenomenon, called refraction; however, once in the second medium, the path was again a rigorously straight line. Here was a demonstration of light gradually curving in the parabolic arch formed by the descending water stream. You will eventually see how this behavior was reconciled with the known properties of light; however, it caused a certain amount of debate at the time. This simple and striking experiment is easily performed—you might want to try it.

THE PHOTOPHONE

In the 1880s Alexander Graham Bell did a considerable number of experiments with a device he called the photophone. Bell personally considered that the photophone was a greater invention than the telephone because it did not require wires to connect the transmitter and receiver. Unfortunately, the photophone was ahead of its time in that the technology to support it was not to become available for some years. Before the invention of the electric light, Bell had to use sunshine, redirected by mirrors to illuminate the diaphragm that performed the modulation function in the photophone. For receiving, Bell used a selenium resistor whose resistance varied with the light intensity. The selenium resistor was mounted in a parabolic reflector which concentrated the light and was connected to a battery and a telephone receiver. All these were wired in series. As one spoke into the speaking tube, the mirrored diaphragm vibrated from the sound waves, thereby modulating the strength of the light waves. The varying light waves in turn varied the resistance of the selenium resistor, which in turn varied the current through the headphone. The device did work, but not easily enough, well enough, or over a sufficient distance to permit it the success attained by the telephone.

The heliograph, or sunlight telegraph, and the blinker searchlight were used by the larger armies and navies of the world to telegraph messages without wires; however, relatively little was done about practical voice transmission by light waves except for the *Lichtsprecher* used by the German navy during World War II. By 1910, Hondros and Debeye had published studies on dielectric guiding of waves from a theoretical standpoint. Other theoretical studies of light transmission were performed in the 1920s and 1930s, but none of these held the appeal that went with microwave transmission by waveguide, which made radar possible.

THE FIBERSCOPE

In a practical sense, this field lay fallow until the 1950s when VanHeel, Hopkins, and Kapany developed the flexible fiberscope.

During this work Kapany developed the first practical glass-coated glass fiber and coined the term "fiber optics." The fiberscope was an immensely useful device. In this device, a carefully sorted rope of parallel fibers used a short focal length lens to conduct the image formed on one polished end of the rope back to the eyepiece, where the image seemed to be in the plane of the fiber. This was magnified by the eyepiece and collimated for viewing. The remarkable part about this was its flexibility. The device would permit one to see around corners and curved obstacles. The image was not even disturbed when the scope was tied in several knots!

The use of the fiberscope for the inspection of welds inside reactor vessels and combustion chambers for jet aircraft engines can hardly be exaggerated. Just as significant is the use of the fiberscope in the medical field, where the small size and flexibility of the device made practical the inspection of the esophagus and the digestive tract without surgery. Early instruments suffered perennially from broken fibers, and the field of view always had a few black spots. Also, they were not as flexible as wished. Nevertheless, they represented a remarkable breakthrough in technology.

THE LASER

For communications work, however, some very significant pieces were still lacking. For one thing, there was no practical way of getting any significant power level of light into the tiny fibers. For reasons which will be studied in subsequent chapters, the projection of a reduced image of a large distributed light source such as an incandescent lamp filament does not efficiently do the job, but one of the pieces fell into place in 1958 with the invention of the laser by Townes.

The laser held forth the promise of providing an extremely intense source of coherent monochromatic light of very small physical and angular extent. For the first time, there was hope of being able to load light powers measured in watts into single glass fibers. This was all that was required to set off—perhaps justify before the lab finance committee—a series of experiments into clad glass fibers by K.C. Kao and G.A. Hockham at the Standard Telecommunications Laboratories in England. At the Glass Museum in Corning, NY, in 1958, one could see a 32-foot long bar of optical glass which had been polished at both ends for use in a submarine periscope. If one looked into one end, he could perfectly read a playing card placed at the far end (I believe it was the Jack of Hearts). However, nine years later in 1967, the losses in good fibers were reported at 1000 dB/km.

Losses such as this would make it necessary to have repeaters spaced very close together if any significant distance were to be covered. Suppose that you could launch as much as 0.1 watts into the fiber and could usably detect as little as 0.000000001 watts. The difference between these levels is 80 dB; therefore, the repeaters would have to be spaced at 80-meter (262-foot) intervals. The system would not work over the length of a football field without an expensive repeater.

However in 1970, just three years later, Kapron, Keck, and Maurer at the Corning Glass Works announced a startling breakthrough in the achievement of fibers hundreds of meters long with losses measuring less than 20 dB/km. This improvement was so dramatic that repeaters could now be spaced 4 km apart. Today, practical off-the-shelf fibers with losses in the range of 6 dB/km are available, and premium priced cables running as low as 3 dB/km can be special ordered.

In addition to the improvements in the fibers themselves, various manufacturers are offering complete transmitters and receivers capable of accepting TTL level inputs and providing TTL level outputs at various video bandwidths. A number of connector firms are beginning to offer connectors for attaching the fibers to transmitters and receivers. In the connector field, in particular, it is difficult to resist drawing a parallel to the era of about 1940 when standardized coaxial connectors for possible interchange first became commercially available. The era of the fully characterized connector in fiber optics is still not with us, though. It is not yet possible to purchase an optical BNC or N connector that is fully described with tolerances, reflection coefficient, and guaranteed fit.

It is obvious that such standardization will have to come before the full potential of fiberoptics is met. We will have to be able to purchase the optical equivalent of AN/UG-58-A/U chassis connectors and AN/RG8-U cables before the instruments and the apparatus become much more than a pioneering curiosity. However, the signs are there indicating that the changes are on the way. The great standardization push in microwave devices came from the exigencies of World War II. Early in the war, a piece of UHF or microwave equipment came complete with its cables and its coaxial fittings and it would connect to nothing else in the world.

If someone wanted to develop a new instrument or attachment to add to the assembly, he had to design and machine the piece. By the early 1950s these difficulties had vanished for most of the common applications; only when broadband "toll ticket" waveguides or something equally exotic were necessary did the fittings and instruments still need machining in the model shop.

These phases of development seem to lie in the immediate future. Within the next few years, standard connectors should become available.

SO MUCH FOR HISTORY

Up to this point our discussion has concentrated on the historic origins of fiberoptic devices; however, we have not directed our attention to the question of the chapter title—why fiberoptics? It would seem likely that there would have to be some fairly powerful incentives to prompt so much research and development activity. Why should all these people put all of this effort into the development of expensive and sophisticated communication devices when the communications arts are already highly developed?

In the words of a song which was very popular a few years ago, "Happiness is different things to different people." Much the same thing can be said for fiberoptic devices. The appeal of fiberoptic devices depends upon certain incentives. It can be economic, technological, or even a logistic advantage that glimmers upon the horizon. The fields must be individually addressed for us to be more specific. We may then begin to see why this new field is so exciting.

TELEPHONE COMPANIES

The basic drive toward fiberoptics in the telephone industry stems from two fundamental advantages: logistics and economics. Considering the first point, logistics, this is the feature which would seem to be the closest to realization in the immediate future. A little more than a generation ago the gas companies in most major cities converted from using coal gas to using natural gas despite the then slightly higher cost of natural gas. The natural gas burned cleaner and was less dangerous than the coal gas; however, these were not the driving incentives for the gas companies. The real push came from the fact that natural gas contained about twice as much heating value per cubic foot as coal gas did. The use of gas was on the increase and most major cities had very large fixed plant investments in gas mains. The conversion to natural gas had the effect of doubling the capacity of the fixed plant or gas mains. In short, the gas company could double its service capability without having to dig at all.

A very similar argument applies to the present position of the telephone companies. Most major cities have very large investments in fixed plants with conduits beneath the streets. The demand for telephone service continues to grow at a relatively steady rate on the order of five percent per year. The real lure of fiberoptics is the fact that size-for-size, a fiberoptic cable using present technology will handle something like 20 times the traffic that can be handled on copper circuits. Foreseeable advances could boost the advantage by factors of 10 or even a hundred.

The cost per foot of the cable itself is becoming roughly competitive with the copper cable cost, although the terminal equipment is very much more expensive. The overriding advantage of being able to expand the system capability to meet future needs without digging is the motivating force in the drive toward fiberoptic telephone trunks. The logistic advantages are sufficient to override some amount of immediate economic disadvantage in the equipment itself.

In economic terms, the advantages of fiberoptics lie in the future. The expense of the required terminal equipment for fiberoptic transmission will probably decrease fairly rapidly on a "learning curve," which is typical of solid-state devices. The prices will probably not come down as fast as those of pocket calculators did; however, a price descent and function growth comparable to the microprocessor or semiconductor memory would not be surprising.

For the cables themselves, a very rapid cost descent seems quite likely. Present telephone trunk cables have a price which is directly related to the cost of electrical grade copper; in fact, the raw copper cost makes up about half of the cost of the cable. Fiberoptic cables, on the other hand, have an active element, the glass or plastic fiber. Glass is among the most inexpensive substances known to man. The treatment required to turn ordinary sand into a fiberoptic strand is an extremely sophisticated process requiring some very expensive machinery. However, a little bit—a very little bit—of this precious treated glass goes a long way. A three-inch diameter and 18-inch long boule or billet would produce 42.4 miles of 200-micrometer fiber and 6.6 miles of 500-micrometer fiber. In a long cable, the fiber itself contributes a nearly negligible portion of the material by weight.

The major portions of the cable material consist of plastics and perhaps reinforcing steel tensile members. Compared to a copper telephone cable of the same information capacity, these materials are used in much smaller quantity because the cable itself is smaller. Far fewer pounds of material are required per mile of cable.

If the glass fiber drawing operation follows the history of the silicon crystal growing and sawing operation, expect the cost of the fibers themselves to drop rapidly as the delivery of large volumes of fiber acts to dilute the capital plant costs and research and development costs. Like the silicon growing operation, the fiber doping and drawing operation lends itself to very high levels of automation. Such operations tend to be very expensive at the outset and eventually become very inexpensive. A third net benefit stems from the reduction of shipping and handling expenses due to the lighter and smaller glass fiber cables compared with copper cables of the same information capacity.

The summation of these three factors holds the promise of very substantial per mile cable cost reductions for fiberoptic cables as cable usage becomes more widespread. If the present installations prove to be satisfactory in operation, fiberoptic installations will grow in number because of first, logistic considerations, and eventually, economic considerations.

MEDICINE

One of the first uses of fiberoptic devices was in the fiberscope, which permits internal examination without surgery or X-ray techniques. This usage is likely to continue and grow. Current fiberscopes are greatly improved, with multiple viewing, biopsy, and manipulator functions added.

An additional area also seems likely to experience a substantial growth. Seriously ill patients are usually monitored for a number of vital signs: electrocardiogram, electroencephalogram, blood pressure, respiration rate, temperature, etc. In particular, two things are troublesome about the existing arrangement.

First, great care must be taken to electrically isolate the sensors to prevent electrical shock. Shocks forcing currents as small as a few milliamperes can be lethal to seriously ill patients. If applied directly to the heart, as could be the case with an implanted pacemaker or EKG electrode, this could be very likely fatal.

Secondly, there is the sheer complexity of the maze of wires to be attached to the patient. A small battery-operated unit serving as a data gathering front end and a multiplexer could be easily arranged to take all of the monitor signals over a single, shockproof, fiberoptic waveguide, which has no metallic component to transmit dangerous electricity.

In a hospital, data from the individual intensive care units could be carried to the monitors at the nurses' station with minimal space and complete freedom from interference caused by X-ray machines, cautery, fluorescent lights, elevators, suctions, and other appliances.

INDUSTRY

In industry, the increasing use of automation and monitoring sophistication has served to multiply the number of wires and cables strung around plants and factories. In these cases, using data multiplexing on fiberoptic waveguides frees a company from the complexity and confusion of existing wiring. It also offers complete freedom from electromagnetic interference from electric welders, lighting, and electrical machinery. The system is more compact, easier to maintain, and far more reliable than other techniques presently permit.

AIRCRAFT

The nature of aircraft is such that many dollars can be spent on one aircraft design if the effort will reduce the weight by even a single ounce or the maintenance downtime by only a few minutes each month. For this reason, fiberoptic transmission systems have been incorporated into a number of current aircraft for the monitoring of remote sensors, remote control, etc. These applications are expected to continue to grow. The weight savings and reduction in cabling complexity (with resultant maintenance reductions) are more than sufficient to offset the price disadvantages.

CABLE TV

For the cable TV industry the promise of fiberoptics is a bit more distant than it is elsewhere. Why? It depends upon a more favorable economic picture in terms of the cost of the cables themselves and the multiplexing and demultiplexing equipment. However, a golden promise it is. The nearly unlimited bandwidth implies that subscriber services, such as direct memory access from home computers to central data banks, could be offered. Leased video connections between plants and downtown offices is another lucrative possibility. As long as the cable TV firm has the franchise to install its cables on the poles and the requirement to maintain them, they might as well sell as much in the way of services as the market will support.

In addition to economic considerations, cable TV faces the hurdle of obtaining from the Federal Communications Commission suitable definitions of the services it may offer. Many of the possible two-way services that are technically feasible lie within areas that could be interpreted as telephone services and are therefore within the province of telephone companies.

COMPUTERS

In the computer area, the broad bandwidth and freedom from interference that a fiberoptic link can offer would make possible some substantial changes in the organization of computer systems. For example, consider one of the 16-bit microprocessors, such as the Motorola 68000. This unit has 23 address lines and 16 data lines in addition to a variety of control lines. These lines are essentially asserted in parallel at a data rate on the other of 4 MHz. In serial format, multiplexed onto a fiberoptic link, a 50-bit word at a baud rate of $50 \times 4 \times 10^6$, 2×10^8, would be required for a real-time direct memory access. This would require a video bandwidth of at least 6×10^8 Hz; however, consider the computing power available with such an arrangement.

In an engineering office, each desk top computer could have the power and data bank resources of the corporate main frame—all in real time. By comparison, about the fastest link available today operates at 9600 baud over a leased telephone line. The data transfer rate would be increased by a factor of 41,667! The pause between the time that the clerk types in your request for an airplane seat and the response would essentially disappear.

The ability to access the main memory bank in real time would bring about a revolutionary restructuring in the distribution of computing power in most major installations. This change would probably increase the power and versatility of the remote terminals. It is very likely that the central computing facility would regress to serving only the function of a central data bank and a system monitor. No longer would it be necessary to have a main computing facility with a few remote terminals. The intelligence of any system would be scattered throughout the system for use where needed, and any terminal could have real-time access to multiple distributed data banks.

AND ELSEWHERE

For a number of years, certain automobiles have used fiberoptic strands to monitor the operation of all outside body lights. Fiberscope monitors to permit the internal viewing and monitoring of nuclear power reactors seems to be in the offing. Nearly "tapproof" telephone communications are possible over fiberoptic links. The future uses for fiberoptic devices seem limited mainly by the imagination of engineers and scientists and the cost factor.

2

Modulation Techniques

In the following chapters we shall make reference to terms such as bandwidth, phasing, and modulation and mention the importance of such things as phase shift. While these concepts are familiar to radio and communications engineers, they are somewhat foreign to many digital people. Therefore, a little consideration of some of these concepts follows.

This discussion is particularly important because the principal work that most engineers will perform with a fiberoptic system will concern itself with the method of modulating the source, the method of decoding the source modulation at the far end or receiving end of the system, and the circuitry involved in translating signals to and from the outside world into the fiberoptic system. In general, the design and development of the light sources and detectors themselves and the fibers themselves is performed by a handful of people working in very specialized laboratories.

As will be shown later, the actual construction of a light-emitting diode or an injection laser diode or a photo transistor is an extremely painstaking process that requires hugh capital investments in vacuum furnaces, vacuum deposition equipment, and micromanipulation equipment. In fact, the portions of these devices which actually do the work are often so small that they can only be seen with the aid of a scanning electron microscope. For this reason, the number of people working directly on the development of the devices is very small and will probably remain so.

The overwhelming majority of the people who will work with

fiberoptic systems will simply select and purchase the light sources and the fibers themselves, as a matter of simple economic necessity. To perform an intelligent selection, it is important to understand the nature and principles of these devices. This is the topic of the subsequent sections of the book; however, the largest portion of the engineering involved in the use of fiberoptic systems will remain in the areas of devising the circuitry and mechanism for translating the information to be transported into light signals, and then translating the light signals back into the original information form. This chapter will introduce some of the concepts of these techniques.

MODULATION

For any form of intelligence to be conveyed over any form of communication channel, it is necessary that some change be made in the operation of the channel. This change is termed *modulation*. It is the modulation which conveys the message. For example, when the telegraph was discussed in Chapter 1, it was noted that the telegraph consisted of a conductive path which could be a pair of electric wires or a single electric wire using the earth for a return conductor. If a battery is applied at the near end of the wire, someone at the far end with a suitable instrument can determine that a current is flowing. The device for detection can be as simple as a lamp bulb. When the bulb is lighted, it can convey one message; when the bulb is extinguished, it can have another message. The lamp which indicates the oil pressure in your automobile is that sort of arrangement. If the bulb lights, it indicates that the oil pressure in the engine is not satisfactory. If the bulb is out, it indicates satisfactory oil pressure.

The simple nature of this communication can be seen to be relatively limited. The system has gone to the expense of providing a source, a communication path of wires, and a detector (the bulb), and it still can convey only a simple yes/no message. Obviously a system long enough to qualify as a communications link would be impractical if it were so confined. Prior to the Morse telegraph, a number of workers in Europe and England constructed telegraphs in which messages could be sent by energizing the correct wire of an assembly. With a Wheatstone telegraph, the sending operator would close a switch corresponding to the letter to be sent. This would energize an electromagnet at the receiving end. A compass needle would in turn swing around to point at the energized magnet, which was labeled with the appropriate letter. This system was actually used in England for a number of years, but it had some very substantial disadvantages.

First, the system required 26 wires for the letters of the alphabet, 10 for the numerals, and a few for punctuation and spaces. The cost of the wire itself was substantial, so the system was very expensive to construct.

A second difficulty was its slowness. It took some time for the compass needle to swing around and settle down so that the receiving operator could be certain which letter was indicated.

The system was also difficult to maintain. With perhaps 40 wires connecting the sending and receiving end, it was very difficult to establish the system so that the correct wire was connected to the correct magnet. A single wind or sleet storm could knock down some or all of the wires and leave the system in a hopeless mess.

The real contribution of Samuel F.B. Morse was the invention of modulation in the form of the Morse code. With the Morse telegraph, only a single wire was used; the return current flowed through the earth. Each character in the language was assigned a unique pattern of long and short current impulses. Initially, these impulses were used to deflect a pen leftward and rightward. The receiving operator would then decode the zigs and zags on the paper into letters. It was quickly discovered, though, that with a little practice, the operator could decode the message simply by the sound made by the pen. The short pulses with two closely spaced clicks were called dots and the long pulses were called dashes. The system worked much faster than the Wheatstone telegraph, and within just a few years, messages were being sent at 50 words per minute.

Along with the development of the telephone came another advance in the concept of modulation. Alexander Graham Bell was a teacher of the deaf. His many experiments had made him well aware that sound is conveyed by pressure waves in the air. Bell reasoned that if these pressure waves could somehow be translated into a varying strength electrical signal and the electric signal subsequently translated back into pressure waves in the air, he would have a mechanism for conveying messages and information over great distances. Furthermore, the instrument could be used by anyone, not just the few people who had the skill to decode the clickety-clicks of the telegraph. The ability to copy Morse code is a skill roughly comparable with the skill required to play a violin—and was developed by about the same number of people. The telephone could be used by anyone with normal speech and hearing.

Bell eventually succeeded in causing the pressure waves of sound to vary the current in a wire and causing the varying currents to vibrate a metal diaphragm to re-create the pressure waves of sound. It was Thomas Alva Edison who invented the carbon microphone that gave the first really loud and clear telephone transmission, however. In the carbon microphone a cell containing loose carbon granules was attached to a metal diaphragm. When the air pressure was high, the diaphragm squeezed the granules and reduced the resistance of the cell. When the pressure was low, the granules relaxed and the resistance increased. The variation in resistance caused the current in the circuit to vary. The Bell receiver operated using the force of an electromagnet on a metal diaphragm. When the current was strong, the steel diaphragm was strongly attracted and caused to bow toward the magnet. When the current was weak, the diaphragm would relax back toward the unstrained position. This vibration of the diaphragm re-created the pressure waves in the air.

THE CARRIER WAVE

When the early radio experiments were performed, the major interest was in the quasi-optical properties of radio waves. It had been shown as early as 1836 by Joseph Henry and Michael Faraday that it was possible to induce an electrical impulse into a circuit which was not connected in any way to the exciting circuit. In the late 1800s, Heinrich Hertz demonstrated that electromagnetic waves with properties similar to electromagnetic light waves could be generated with special apparatus, and that a strong enough discharge could be made to excite a spark at the terminals of a dipole antenna which was removed from the transmitter by a number of meters. Heinrich Rudolf Hertz and a number of investigators went on to show that these waves, which we today would refer to as microwaves, had properties similar to those of visible light. They could be focused, refracted, and reflected, and they behaved in all of the classical optical ways. The only difference was in the wavelengths, which were in the 10 to 30 cm range, and therefore tens of millions of times as long as light waves.

The first long-distance radio transmissions were to come using wavelengths four orders of magnitude longer than microwaves. Unlike the microwaves investigated by Hertz, the long waves of Marconi and other investigators would follow the curvature of the earth and were not shielded by hills and trees and the natural terrain. The basic concept used in this transmission was the creation of a *carrier wave* that would be radiated from the transmitter and portions recovered by the receiver or receivers. This carrier wave served a

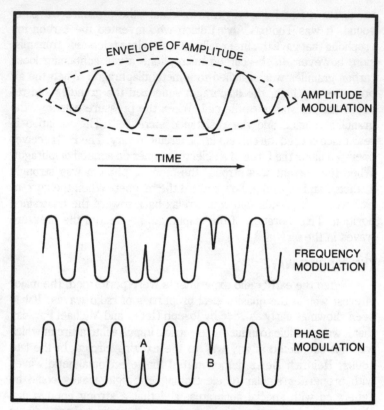

Fig. 2-1. The three fundamental modulation techniques.

function similar to that of the current in telephone lines. It acts as the carrier of the modulation or information.

A fundamental difference in the carrier wave, however, is that it has three mechanisms by which it can be modulated. Figure 2-1 shows these in the order amplitude, frequency and phase modulation. The amplitude modulation at the top shows that the amplitude, or strength, of the signal is varied. In radio telephone, the amplitude is varied about some median level. For example, if the carrier is transmitting the sound of a speaker, a high air pressure in front of the microphone can make the carrier stronger, and a low pressure can make the carrier weaker. The locus describing the peaks of the wave is termed the *modulation envelope*. It is this modulation envelope that carries the information. Amplitude modulation (AM) is the mechanism used for receiving the standard broadcast band. In the extreme case of radio telegraphy and pulse modulation, the signal is turned completely on and off by the modulation function.

The middle waveform shows *frequency modulation,* which alters not the amplitude of the carrier, but rather the frequency. The wave retains a constant amplitude, but the up and down oscillation is much faster at the center than at the ends. In this case, the pressure from the speaker's voice could make the carrier frequency go high when the pressure was high and low when the pressure was low. In the extreme case of frequency modulation, the carrier jumps between two fixed levels of frequency in the analog of the on/off modulation of the telegraph. This is referred to as *frequency shift keying* (FSK).

The technique shown at the bottom of the page is *phase modulation.* In this case, the *phase angle* of the wave is modulated. In telephony, for example, a high pressure at the microphone might cause the phase angle of the wave to advance; a low pressure might cause the phase angle to retard. In the extreme case of phase modulation corresponding to telegraphy, the phase angle jumps between two fixed levels. This technique is known as *phase shift keying* (PSK), and it is shown in the illustration. The phase is completely reversed at points A and B. The extreme case is shown because it is extremely difficult to tell the difference between frequency modulation and phase modulation in normal cases.

The term *phase angle* probably deserves more explanation. Figure 2-2 shows a sine wave, or rather a segment of a sine wave, with the expression for voltage being given by Equation 2-1. The sine wave is described in terms of mechanics as simple harmonic motion. A particle oscillating between any two fixed locations in a fixed period of time experiences the least possible acceleration when traveling in simple harmonic motion. In electrical and acoustical work, the sine wave represents a single pure frequency.

The sine wave represents the height of the tip of the arrow rotating at a constant rate equal to ωt. The term ω is the *angular frequency.* Because there are 2π radians in the circumference of a circle, the angular frequency is larger than the rotation rate "f" by this factor.

Consider the matter of phase angle. Figure 2-3 shows the sine waves generated by two vectors which are rotating at the same speed ωt but with the dotted vector ahead of the solid vector by some angle ϕ. The sine waves are identical except for a translational motion along the time axis; that is, the dotted wave is ahead of the solid one. The dotted wave is described as having a *leading phase*

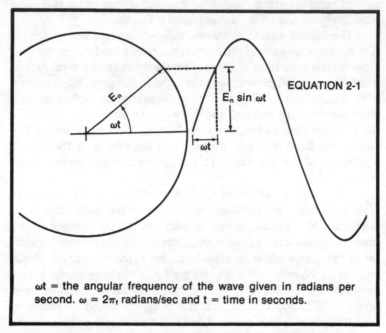

EQUATION 2-1

$E_n \sin \omega t$

ωt = the angular frequency of the wave given in radians per second. $\omega = 2\pi_f$ radians/sec and t = time in seconds.

Fig. 2-2. The height of the rotation vector above the axis describes a sine wave.

angle and the value of this angle is, of course, ϕ radians. It is precisely this angle that is modulated by the modulation function in phase modulation.

What are the similarities between phase modulation and frequency modulation. Let us suppose that the angle ϕ is not a constant but rather that it increases with time; in other words, $\phi = \phi_0 t$. It is fairly easy to see that the dotted vector would get farther and farther ahead of the solid vector until it finally laps the solid vector like a faster race car passes a slower race car several times during a long race. In the time between laps on the slower vector, the faster one would have traced out one extra sine wave; thus, its frequency is higher. Similarly, if the angle ϕ were decreasing with time, the dotted vector would drop behind the solid vector until it was lapped, and the frequency of the solid vector would be higher. In fact, there is an instantaneous difference in frequency at any time that the angle between the vectors is changing. The principal difference between frequency and phase modulation in telephony is that for frequency modulation, a constant fixed pressure on the microphone yields a

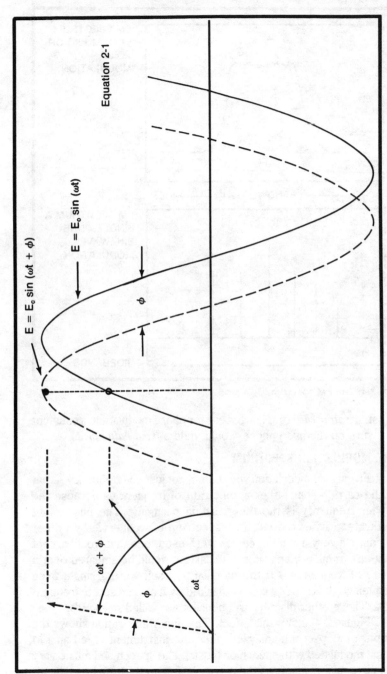

Fig. 2-3. The definition of phase angle.

25

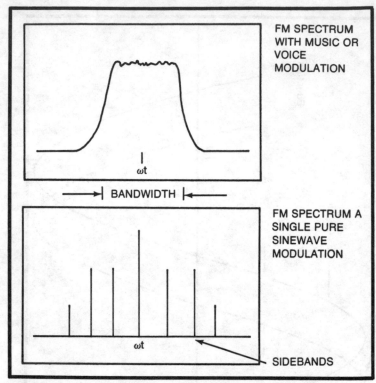

FM SPECTRUM WITH MUSIC OR VOICE MODULATION

ωt

→| BANDWIDTH |←

FM SPECTRUM A SINGLE PURE SINEWAVE MODULATION

ωt

SIDEBANDS

Fig. 2-4. The FM spectrum.

constant *rate of change of* φ whereas for phase modulation, a constant pressure on the microphone would yield a *fixed value of* φ.

THE MODULATION SPECTRUM

Frequency modulation will contain not just one frequency—that of the carrier—but rather a spectrum of frequencies because the carrier frequency is the thing that is changing. The phase shift modulated carrier represents a spectrum of frequencies. In other words, if a very narrow receiver were used, one that would accept only one frequency at a time, the carrier would be smeared over a range of frequencies. If the modulation itself were a single fixed frequency, the carrier would be flanked with other discrete frequencies. These other discrete frequencies are called *sidebands*.

Figure 2-4 shows this effect. The upper illustration shows the output of a spectrum analyzer when the instrument is fed an FM signal modulated with speech or music. The speech and music vary in an irregular manner, so the carrier is smeared over a broad band. The bandwidth is determined by a number of factors including the *modulation index,* the *deviation factor* and the *content of the modulation.* With a single tone, the modulation is not smeared. Instead, fixed *sidebands* exist at certain frequency differences from the carrier.

THE AM SPECTRUM

Although it is relatively easy to see that frequency and phase modulation can smear the carrier, the fact that the spectrum of an amplitude-modulated signal will also be smeared is not so obvious. To understand this, consider the case shown in Fig. 2-5. At the top is an illustration, frozen in time, of a rotating carrier vector that rotates at a frequency $\omega_c t$.

Attached at the top of this are two other vectors that represent two modulation vectors. These rotate with respect to the tip of the carrier arrow with angles $\pm \omega_m t$; that is, they rotate in opposite directions. Assume that the modulation rotation is very slow with respect to the carrier rotation and that the modulation vectors are each only half as high as the carrier vector.

The sequence of sketches beneath the large figure are after a large number of rotations of the carrier arrow. Despite the spread in time between them, each one catches the carrier arrow when it is in the vertical position.

In Fig. 2-5A, the modulation vectors and the carrier vector are all lined up; therefore, they add directly, and the total amplitude is just two times the carrier amplitude. In Fig. 2-5B, which is many rotations of the carrier later, the modulation vectors have rotated $\pm 45°$. The sideways components cancel, so the only effect is caused by the projection of the modulation vectors upon the carrier vector. This will yield an amplitude of $E_c + 0.707E_c/2 + 0.707E_c/2$, or $1.707 E_c$.

In Fig. 2-5C, the two modulation vectors are diametrically opposed and therefore neither add to nor subtract from E_c. In Fig. 2-5D, the modulation vectors are partially canceling E_c and the resulting amplitude is $0.293 E_c$. In Fig. 2-5E, the modulation vectors have completely canceled the carrier.

If there were additional illustrations, you would find that F would be like D and G would be like C and soon until the cycle completed itself with I being identical to A. If locus of the envelope were plotted, it would be a sine wave centered about E_c.

On a spectrum analyzer, there would be just three lines, as shown in Fig. 2-6. The carrier would be present in the center, and it would be flanked by two sidebands having only half the voltage. The sidebands would be displaced in frequency by the modulation fre-

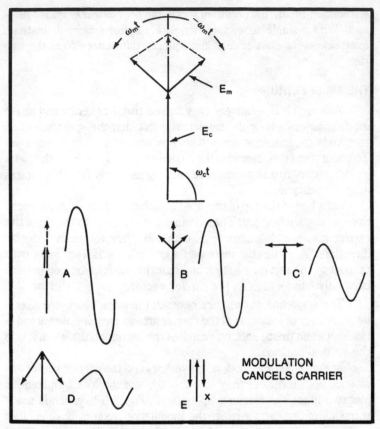

MODULATION
CANCELS CARRIER

Fig. 2-5. Amplitude modulation.

quency from the carrier. Because the power in a signal is proportional to the square of the voltage, each of the sidebands would have one-quarter as much power as the carrier. In sum, they would have one-half as much.

The power distribution is the reason that most high-frequency radios have turned to single sideband operation in recent years. The carrier contains no information whatever, and only one of the sidebands is required for the full transfer of information. The other sideband is redundant and not required. Transmitting both sidebands wastes transmitter power. It also makes the spectrum twice as broad as necessary. In essence, a single sideband radio can convey the same amount of information with one-sixth the power and one-half the spectrum usage.

Figure 2-5 considered only the case where the modulation vectors were precisely half as great as the carrier vector. From Fig.

2-5E, it is apparent that this is the largest that they could be from a practical standpoint; if they were any larger, they would require the generation of negative power at Fig. 2-5E if the signal were not to be distorted. Because this is the largest modulation the carrier can handle without distortion, it is referred to as 100 percent modulation. In actual radio transmission, the modulation is seldom allowed to exceed 85 percent to avoid distortion and spectrum splatter on modulation peaks. With voice or music modulation many frequencies are present, so the spectrum does not consist of just the three lines shown in Fig. 2-6. Rather, it is a dancing smear of sidebands, as shown in the FM picture.

CARRIER COHERENCE

The expression, $E_o \sin \omega t$, is really a mathematical abstraction. No such voltage has ever or will ever exist. Consider for a moment that the function is completely unvarying and eternal. You can calculate the voltage precisely for any point from the beginning of time to the end of time and the peak voltage and the zero crossings repeat themselves precisely in lock step throughout eternity. The function can therefore be described as being fully *coherent;* that is, given an instantaneous reading of the voltage at any point in time, the voltage at any other time can be calculated with perfect precision.

Real sources of carrier waves tend to depart considerably from this behavior. For one thing, they have to be started at some instant.

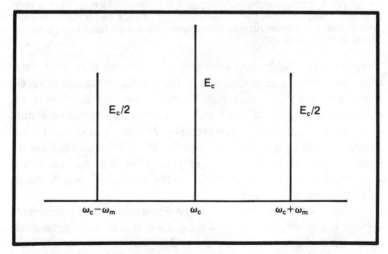

Fig. 2-6. Spectrum for amplitude modulation with the modulating vectors equal to one-half the carrier vector.

29

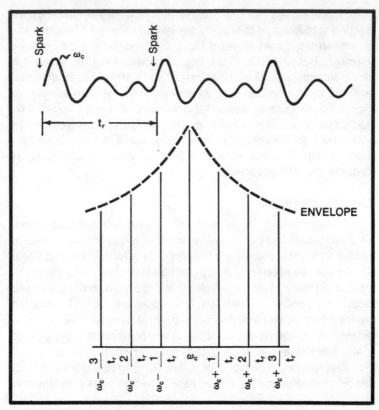

Fig. 2-7. At A, damped wave generation. At B, if the sparks had come at very precisely timed intervals and the new wavetrains were generated precisely in step with the preceding trains, the spectrum would have been a clean series of discrete lines.

They will all probably stop eventually, too. At the very least, this constitutes amplitude modulation, so there is some amount of spectrum spreading in the very finest source; however the degree of coherence that can be attained with such things as a quartz oscillator or a caesium vapor clock are startling. For the quartz clock, the frequency stability is as high as one part in 10^8 which would amount to a gain or loss of 1 second in 3.16 years. For the caesium beam standard, the stability can be as high as one part in 10^{10}, which would be one second in 316 years!

Other factors are involved in the generation of truly coherent signals. In the very early days of radio, radio signals were generated by applying a high voltage to a coil and condenser tank circuit. The voltage would spark across a gap and set the tank circuit to ringing like a bell. When the voltage had built up again, another spark would occur and a new damped wavetrain would begin. This effect is shown in Fig. 2-7. If the spacing in time between the successive sparks had been a precise number of microseconds apart, the wavetrain would have been more or less coherent. In other words, the voltage at any other time could be predicted from a single time sample; however, nature does not give in quite that easily.

In a practical spark radio, the sparks took place in a more or less random order. Therefore, instead of a spectrum made up of a series of neat lines with nothing in between, the spectrum was a rather broad mishmash smeared out over a considerable portion of the spectrum. The signal spectrum was called noisy or incoherent. This characteristic applies as well to light-emitting diodes, laser diodes, and continuous wave gas lasers. This is very important because it severely limits what can be done in the way of modulation of a fiberoptic system. There are no coherent optical generators in the radio sense of the word. A device with a smeary spectrum of the type shown at the bottom of Fig. 2-7 can only be modulated by relatively crude amplitude modulation methods. Something as sophisticated as frequency and phase modulation cannot be applied directly either. And single sideband modulation is completely out of the question at optical frequencies with the existing state of the art.

There are other techniques by which some of these methods find themselves into fiberoptic systems. The most common of these is the *subcarrier* technique. An electronically generated coherent signal is applied to the relatively incoherent optical signal by means of some form of amplitude modulation. The coherent subcarrier can then be modulated in any of the common sophisticated techniques. The advantage of this arrangement is that a large number of such subcarriers can be applied to a single optical signal. After detection of the amplitude modulation on the optical signal, these subcarriers can then be singled out just as your TV or FM radio singles out individual stations from the jumble of electromagnetic radiation in the air.

MULTIPLEXING

The concept just described is generally called *frequency diversity multiplexing* (FDM) because each separate signal on the system is assigned a discrete carrier frequency. The FDM technique is commonly used when many individual complex and high-speed signals such as a number of TV channels are to be applied to a fiberoptic system. The term *multiplexing* itself usually means the use of a single channel to accommodate a multiplicity of independent signals.

Transmission of multiple TV channels over a single fiberoptic link is only one of many examples of a multiplex operation. Obviously, a CATV distribution network would enjoy economic benefits from fitting all of the different signals upon a single fiber.

A similar comparison applies to telephone trunks where the current state of the art places 680 telephone conversations upon a single fiber. In less sophisticated industrial fiber links, the same principle applies. It is usually advantageous to send the readings from a number of instruments over the same fiber. The cost of transmitter, receiver, and fiber is such that it is generally not economical to use the fiber for only a single communication channel.

The multiple carrier FDM system is by no means the only technique for applying multiple information streams to the same channel, nor is it necessarily the best for all situations. Another common technique is *time division multiplexing* (TDM). In this technique, there is generally some synchronizing pulse to inform the receiver that the transmission sequence is about to begin. Thereafter, each separate input channel is assigned a discrete time slot and its message is transmitted during that time slot. At the receiving end, the time slots are counted and the data for a specific instrument can be separated out or the readings of all of the instruments can be reassembled.

The relative advantages of FDM stem from the fact that each input signal is always available. For a wideband, rapidly changing signal, such as a TV signal, this is a prime requirement since one could not afford to lose any significant portion of the picture or a single frame. The principal disadvantage of FDM is that there is always some finite limit on the available power from the transmitter. Because the voltages from the individual signals can add at some time, this generally means that the power must be divided as follows:

$$P_{max} = (Ne)^2$$

$$= N^2 e^2$$

Thus: $$P_{max}/e^2 = N^2$$

Simply analyzed this means that the maximum power available for any one channel in an FDM system is inversely proportional to the square of the number of channels. Doubling the number of channels will reduce the available power to any one channel by a factor of four. Without this limitation, there will be times when the sum of the

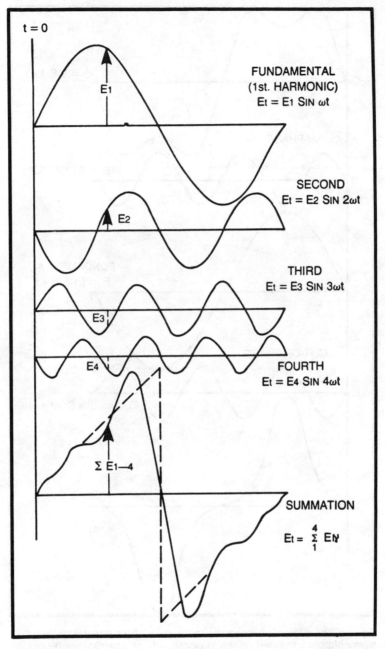

t = 0

FUNDAMENTAL
(1st. HARMONIC)
$E_t = E_1 \sin \omega t$

SECOND
$E_t = E_2 \sin 2\omega t$

THIRD
$E_t = E_3 \sin 3\omega t$

FOURTH
$E_t = E_4 \sin 4\omega t$

ΣE_{1-4}

SUMMATION

$E_t = \sum_{1}^{4} E_N$

Fig. 2-8. The fourth harmonic resultant provides a good approximation of a sawtooth wave.

33

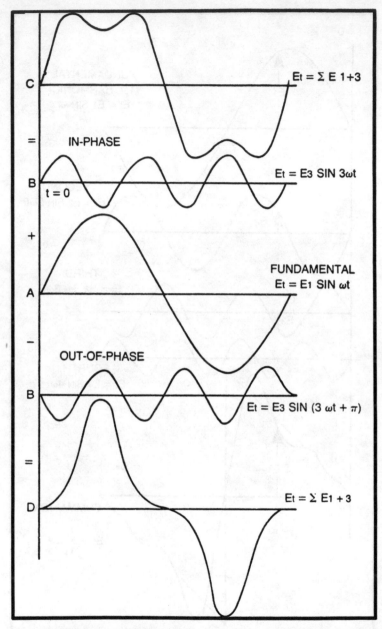

C ————— $E_t = \Sigma\ E\ 1+3$

=

IN-PHASE

B ————— $E_t = E_3\ SIN\ 3\omega t$
t = 0

+

FUNDAMENTAL
$E_t = E_1\ SIN\ \omega t$

A —————

−

OUT-OF-PHASE

B ————— $E_t = E_3\ SIN\ (3\ \omega t + \pi)$

=

D ————— $E_t = \Sigma\ E_1 + 3$

Fig. 2-9. The effect of phase relationship when a fundamental and the third harmonic are combined. When in phase, a near square wave results, as shown at the top. When out of phase, the resultant is sharply peaked, as at the bottom, resembling a sin² function.

34

powers will exceed the capacity of the system and the full information will not be sent for each channel. Sophisticated techniques can ensure that the peaks do not take place simultaneously on all channels; however, in an FM or PM channel, the power is constant and the N^2 limitation applies.

On the other hand, the TDM channel devotes the full power of the system to each individual channel but only for a fraction, which is something less than 1/N of the time. In terms of energy delivered per channel (energy = power × time), the TDM system has a net advantage; however, the sampling rate must be faster than the fastest variation that must be reproduced in any channel. In a TV picture, the maximum video frequency is about 3.5 MHz. To have a reasonable reproduction it would therefore be necessary that the sampling take place something like three times per cycle. The sampling rate for TV video would be such that the sampling must take place 3 × 3.5 × 10^6 × N times per second. If very many TV channels were sent, the sampling rate would become very high.

Of course, the data is changing quite slowly in a great many

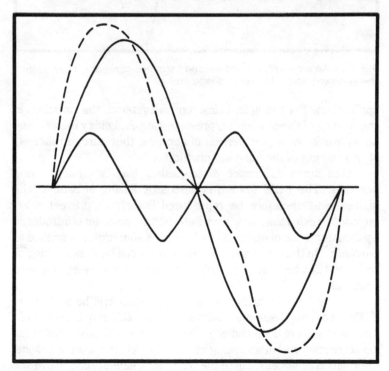

Fig. 2-10. The fundamental plus the second harmonic.

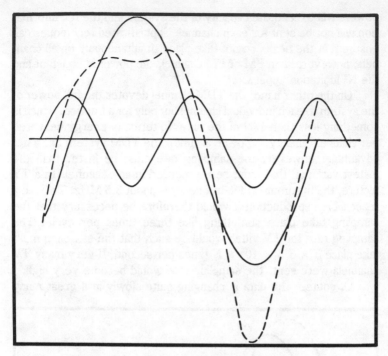

Fig. 2-11. An arbitrarily phased second harmonic component can destroy the symmetry about the zero voltage axis.

applications. For example, in instrument systems, the variations in the reading of thermometers, pressure gauges, acidity meters, and so on usually vary over periods of seconds; therefore, a relatively slow scanning of the inputs is practical.

One more difference distinguishes the two multiplexing schemes. The FDM system is inherently analog in nature. The signals must therefore be reproduced linearly to prevent *intermodulation distortion,* which is a distortion in one channel introduced by the interaction of other channels. This is sometimes described as *crosstalk.* On the other hand, TDM systems can be either analog or linear and can be relatively easily arranged to have essentially zero crosstalk.

In general, the highest data rate systems will be analog and FDM since this eases the modulation bandwidth requirements. At any given state of the technology, a wider information bandwidth can be accommodated in an analog modulation scheme than in a digital information scheme. Despite the other advantages, then, FDM will be selected for the highest data rate systems over TDM.

PHASE SHIFT EFFECTS

Any form of modulation or wave shaping has to introduce a broadening of the carrier. Furthermore, many schemes such as pulse modulation introduce a spectrum of sidebands. We shall see shortly that there is a certain amount of dispersion in most fiberoptic systems. This means that components of different frequencies do not propagate at precisely the same speed through the fiber. When the harmonics suffer a phase shift with respect to the fundamental, severe reshaping of the wave can take place.

Figure 2-8 shows how the first, second, third, and fourth harmonics of a wave can add up to yield a fair approximation of a sawtooth wave when combined in the correct proportions and phase. Figure 2-9 shows that with one phasing, the addition of a proper amount of third harmonic can provide a fair approximation of a square wave. When the phase of the harmonic term is reversed, however, the waveshape can approach a $\sin^2$ function.

Figure 2-10 shows still another property. With proper phasing, the presence of an even-order (in this case, second) harmonic component can yield a symmetrical wave; however, the phase shift of the harmonic can destroy the symmetry of the wave about the zero voltage axis.

3

Basic Information Transmission

In this chapter we shall address some of the fundamental aspects of communications engineering with attention devoted to the manner in which they impact upon fiberoptics. To someone familiar with high-frequency radio communications or someone familiar with telephone communications, there will be some rather unfamiliar concepts involved. However, in the main, some of the more basic concepts are transferable from one technology to the other. For the person whose background is principally digital electronics, nearly all of the concepts may be unfamiliar. Accordingly, we shall attempt to make the discussion here basic enough to act as an introduction for those who have not worked in communications.

ATTENUATION

In all forms of communication, there is some mechanism for the loss of signal power with distance and some relationship between the power of the original signal and the power as received. There is also always some relationship between the intelligibility of the received signal and the power level of that signal. We shall first of all concern ourselves with the diminution of the signal over a given distance.

To begin with, consider the illustration of Fig. 2-1, which shows that a source of energy is at the center of a sphere. Presume that this source scatters energy equally in all directions, like a star hanging in space. Now if we consider the energy incident upon a unit area designated as A_e, the amount of the source energy P_t that would strike A_e is proportional to the ratio between A_e and the total area of the sphere.

Nothing about this discussion pertains to anything except the geometry of the situation. The source at the center of the sphere could be a paint sprayer and the statements would apply equally to the amount of paint captured by the unit area. The amount captured by A_e decreases as the square of the radius simply because the area of the sphere grows as the square of the radius and A_e becomes a progressively smaller fraction of the whole.

If the source at the center of the sphere does not spread the energy equally in all directions but rather concentrates it so that more goes in one direction than in others, the ratio between the transmitted and the received power is modified by a factor known as the *gain* of the transmitting radiator, designated by G_t. In this form, the formula is known to radio engineers as *Friis transmission formula*. If the source scatters the energy equally in all directions, $G_\theta = 1$, and the source is known as an *isotropic radiator*. An incoherent thermal radiator such as a star can be an isotropic radiator; however, a coherent, polarized source can never be an isotropic radiator. We shall shortly be discussing the concept of coherence. For a coherent, polarized radiator, G_θ is always greater than unity. G_θ is a geometric parameter related only to the ratio of the power scattered in the direction where the radiation is strongest to the power which would have been scattered in that direction if the radiator had scattered the energy isotropically.

Now the radiating source may not always manage to radiate all of the energy supplied to it in the form supplied. It can sometimes degrade a portion of the energy supplied to it into heat or transmute it in some other manner, therefore, the lowest formula shows an allowance for this degradation designated by the symbol k. The symbol k is used here to designate the fraction of the energy delivered by the source (P_t) that is eventually radiated by the radiator. The symbol G_θ is the *directivity* of the radiator.

Ear Response

In his studies with the deaf and in subsequent experiments while developing the telephone, Alexander Graham Bell noted that the response of the human ear is logarithmic rather than linear. Accordingly, he applied a logarithmic scale to the measurement of sound intensity and subsequently to the measurement of the power present in the signal on an electric telephone line. This measurement scale was adapted and persists to this day in communications work, although in a somewhat modified form. The ratio between the transmitted and the received powers for the illustration in Fig. 2-1 would be given by:

39

$$\text{PATH LOSS} = 10 \log_{10} \frac{P_t}{P_r} \quad \text{in decibels} \quad \text{Equation 3-1}$$

The modification is the factor of 10 which changes the unit from the bel to the decibel, which is usually abbreviated dB.

In terms of sound, one decibel is about the smallest change in sound level which can be perceived by the human ear. Three decibels represents a change in power level by a factor of two and 10 decibels represents a change by a factor of 10.

If the inverse ratio in Equation 3-1 is used to calculate P_r/P_t, the quantity would be called *path gain*. Because the ratio would have been less than unity, the logarithm would have been negative but numerically equal to the first result.

Consider the relationship in Equation 3-2 of Fig. 3-1 and examine it to determine the manner in which path loss varies with the distance:

$$P_{r1} = \frac{P_t G_t A_e}{4 \pi R_1^2}$$

$$P_{R2} = \frac{P_t G_t A_e}{4 \pi R_2^2}$$

$$\frac{P_{R1}}{P_{R2}} = \frac{R_2^2}{R_1^2}$$

and   Path loss variation $= 10 \log_{10} \dfrac{R_2^2}{R_1^2}$ Equation 3-2

if $R_2 = 2_{R1}$

$$\text{PLV} = 10 \log_{10} \frac{4 R_1^2}{R_1^2} = 6.021 \text{ dB}$$

This spreading loss is equal to 6 dB whenever the range doubles. Stated differently, a doubling of the range reduces the captured power by a factor of four, which is equal to 6 dB.

Rigorously speaking, the decibel scale should only be applied to power; however, voltage measurements are frequently stated in decibels. This follows from Ohm's law:

Ohm's law $\qquad\qquad E = IR$

Power $\qquad\qquad P = I^2R$

Therefore: $\qquad\qquad P = E^2/R$

$$\text{Gain} \;=\; 10 \log_{10} \dfrac{\left(\dfrac{E_1{}^2}{R}\right)}{\left(\dfrac{E_2{}^2}{R}\right)} \qquad \text{Equation 3-3}$$

If the two R's are *identical* they may be canceled to obtain:

$$\text{Gain} \;=\; 10 \log_{10} \dfrac{E_1{}^2}{E_2{}^2}$$

$$\;=\; 20 \log_{10} \dfrac{E_1}{E_2} \qquad \text{Equation 3-4}$$

Since $\qquad\qquad \log_{10} N^a \;=\; a \log_{10} N$

This relationship applies only when both voltages are measured across the same value of R. There is no such thing as a *voltage* dB or a *power* dB. A dB is just a dB.

If the voltages are measured across different resistances, some very peculiar and *inaccurate* conclusions might be drawn. For example, old 78 rpm phonographs used to develop about 1 volt across a million ohms at the input and deliver about 1 volt across 4 ohms at the output. The voltage gain is unity—the same voltage is on the output as on the input. The power gain of the amplifier, however, was one-quarter watt divided by one microwatt, or 250,000 times! Stating the voltage gain in dB without accounting for the vast difference in the resistance would certainly give a very misleading impression about what the amplifier was doing.

Inverse Square Loss

The spreading or *inverse square loss* caused by the spreading of the energy in space is not the only type of loss encountered in

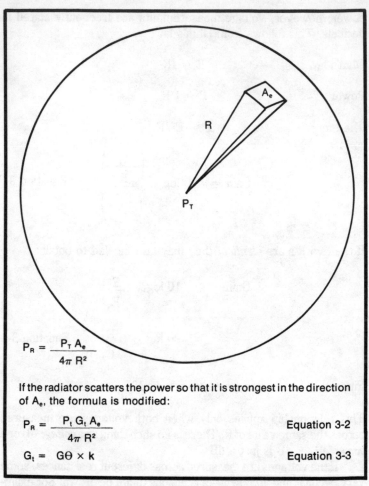

$$P_R = \frac{P_T A_e}{4\pi R^2}$$

If the radiator scatters the power so that it is strongest in the direction of A_e, the formula is modified:

$$P_R = \frac{P_t G_t A_e}{4\pi R^2} \qquad \text{Equation 3-2}$$

$$G_t = G\theta \times k \qquad \text{Equation 3-3}$$

Fig. 3-1. The energy source, P_t, at the center of the sphere scatters its energy equally in all directions. The amount of the energy incident upon A_e is proportional to the ratio of the area of A_e to the area of the whole sphere.

transmission systems. Obviously, if the energy could be completely prevented from spreading, there would be no loss at all if none of the energy were degraded to heat. In empty space it is impossible to prevent the spreading. In a constrained medium such as a fiberoptic waveguide, however, it is possible to prevent spreading. Therefore, if the medium had no losses—that is, if none of the energy escaped and none was degraded to heat—the energy could be sent entirely without loss.

Some losses are always in a network. A reasonable simulation of the loss mechanism can be presented with the lossy transmission line circuit shown in Fig. 3-2. In this network, series elements and shunt elements contribute to the losses. Starting at the load end at point A the 50-ohm load and the 50-ohm loss combine to 100 ohms, which is connected in parallel to the 100-ohm shunt loss element of the cell. This parallel combination again yields 50 ohms, so the impedance iterates through N cells to the source. The impedance looking into any cell is always 50 ohms, and the source sees 50 ohms regardless of the length of the ladder.

In the arrangement shown, every cell starting from the source halves the voltage applied to the next cell. Therefore, an N cell ladder would supply the load with a voltage of $E/2^N$. Stated in decibels, the ladder attenuates N times 6.021 dB.

This attenuation is quite different from the spreading attenuation. For the ladder attenuator, the attenuation is simply N times the attenuation of a given cell; therefore, it is possible to state the attenuation in terms of dB/km or dB/mi. In contrast, the spreading attenuation might be 6 dB from mile 1 to mile 2 and another 6 dB from mile 2 to mile 4 and another 6 dB from mile 4 to mile 8, and so on.

In the example, the rather high attenuation for the cell is not a limiting factor on the structure. For example, if the load resistance remained at 50 ohms and the series resistor were 5 ohms, a shunt resistance of 550 ohms would provide the same characteristic impedance of 50 ohms to the source, but the attenuation would be reduced to 0.828 dB/cell. With a 0.5-ohm series resistance and a 5050-ohm shunt resistance, the attenuation would reduce to 0.0864 dB/cell.

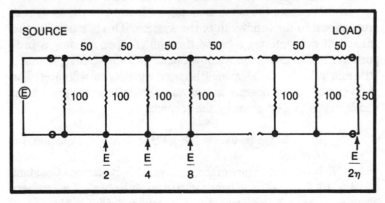

Fig. 3-2. The ladder attenuator.

In electrical engineering, it used to be relatively common to see attenuations specified in terms of nepers, where the neper was defined as:

$$\text{Loss} = \ln \frac{P_1}{P_2} \text{ nepers} \qquad \text{Equation 3-5}$$

This usage is less common today and is seldom seen in any papers on fiberoptics. The term *ln* describes natural base logs or logs to the base *e*. One neper equals 8.686 dB.

NOISE AND DETECTABILITY

The limit of detectability on signals is nearly always given by the noise in the system. In electrical or communications engineering, *noise* is any unwanted or interfering signal that can mask, obscure, or otherwise limit the detection of the desired signal.

In radio systems, noise can arise from natural disturbances such as thunderstorms or atmospheric electricity. It can also be man-made, arising from inadequately shielded ignition systems on gasoline engines, electrical arcs from motor brushes, welding machines, switches or from gaseous discharge in neon and fluorescent lamps. The noise can also originate within the front end stage of the receiving apparatus. In the microwave regions, there is very little atmospheric noise and relatively little man-made noise; therefore, the smallest discernible noise is usually determined by the front end or internal noise of the receiver. In the standard broadcast band (550 to 1550 KHz), the situation is reversed, and the minimum discernible signal is usually limited by natural or man-made static.

In either case, the amount of noise admitted by the system is proportional to the bandwidth of the system. This is much like the amount of rain entering a house through an open window is proportional to the size of the window opening. If the opening is small, little rain will enter; if the opening is large, a great deal will enter. For radio circuits which operate in low-noise environments, the critical sensitivity is usually given by the formula:

$$\text{Noise power} = \text{NF kTB in watts} \qquad \text{Equation 3-6}$$

where NF is the noise figure of the receiver, k is Boltzmans Constant (1.38×10^{-23} watts/cycle bandwidth/degree Kelvin), T is temperature in degrees Kelvin, and B is system bandwidth in Hz.

The noise figure is an experimentally determined quantity that depends upon the system design, the front end components, the front end circuit gain, etc. The Kelvin temperature is simply the temperature of the device in degrees Celcius plus 273°. This is often assumed to be 290° K for most measurements. A good transistor front end on a radio receiver will generally have a noise figure on the order of 3 to 6 dB (a factor of 2 or 4). Very low-noise maser or parametric amplifier front ends will have even better noise figures. Because these devices are often cooled with liquid nitrogen, or even liquid helium, the NF term is frequently omitted and the receiver sensitivity is specified in terms of equivalent noise temperature. In satellite receivers and radio astronomy receivers, the equivalent noise temperature can be 10° K or less.

While the relationship of Equation 3-6 is applicable to very high-frequency devices, semiconductors tend to show excess noise above the thermal noise due to shot effects and surface recombination effects; therefore, the noise from an amplifier whose bandwidth extends to nearly DC is often given in terms of the *noise equivalent power* (NEP), which is specified in watts per $\sqrt{\text{hertz}}$. Typical figures range from 10^{-13} to 10^{-14} for devices suitable for fiberoptic detector service. One of the reasons for the change from linear to square root bandwidth is the presence of significant levels of junction capacitance that tend to short out some of the higher frequency components. The power level NEP must in general be exceeded or at least equaled if the signal is to be detected. The ratio of the signal power incident upon the detector to the internal noise power in a fiberoptic system is called the *signal-to-noise ratio* (SNR). The error rate, or the rate at which the detector mistakes an 0 for a 1, is related to the SNR. In digital transmission, as the SNR decreases, the system begins to make more and more errors until an unacceptable performance level is reached. In an analog system, the falling SNR reduces the quality and intelligibility of the system.

SYSTEM BANDWIDTH

The actual bandwidth occupied by an information stream is a function of a number of things. First and foremost among these is the rate at which the finished information is being transferred. In the ordinary telephone system, the mechanism that transfers speech over the system is, in fact, a form of single sideband modulation in which the carrier frequency is zero or direct current. On the telephone, the voice signals are modulated upon the direct current, which flows at all times. The frequencies between about 300 Hz and

3000 Hz are sufficient to handle all of the information transfer, and in fact to permit voice recognition at the far end. It is not too difficult to understand what the other party is saying on the telephone. Frequently, the speaker can be recognized without clues other than speech and voice characteristics.

Single Sideband

The single sideband, amplitude-modulated signal has been shown to be about an optimally efficient modulation technique when coherent detection is used. In optical techniques, however, the coherent detection of narrowband signals is not very practical. The usual optical detector is not some form of superheterodyne that translates the signal down to intermediate frequencies where ordinary electrical amplification techniques are used. Instead, the optical detector is usually a photodiode, a phototransistor, or an avalanche diode detector. In effect, this is like the simple crystal set in which the received signal is merely rectified and then sometimes followed by amplification. The reason for this is relatively obvious.

Consider a typical radio single sideband signal. The carrier frequency is commonly below 30 MHz. For a single sideband speech signal to be intelligible, the reference carrier that injects the carrier signal required for SSB detection must usually be within 30 Hz of the missing carrier of the transmitter. In this case, this works out to an accuracy of tuning of one part in 10^6. This is not the easiest thing in the world to manage with temperature variation and aging; however, modern military radios are capable of such accuracy and stability.

On the other hand, for a typical operating wavelength of 980×10^{-9}m for a fiberoptic waveguide, the frequency is approximately 3.06×10^{16} Hz. For an accuracy of tuning of $\pm$ 30 Hz, this implies a stability and accuracy of one in 10^{15} which is very close to the limit of man's ability to measure frequency. A single sideband transmitter capable of optically modulating a voice channel in only 3 kHz bandwidth is a goal yet to be achieved. The receiver which will detect it also lies in the future.

As a benchmark, a Hewlett-Packard model 5061 caesium beam frequency standard has a rated settability of $\pm$ 7×10^{-13} in frequency and a short term stability of $\pm$ 1.5×10^{-10} in frequency. This is probably the most accurate frequency source available commercially, and yet it falls short of the stability requirements for coherent SSB modulation by three to five orders of magnitude. Consequently, it seems likely that direct coherent modulation of an optical carrier by SSB, low-index frequency modulation and low-index frequency shift keying (FSK) or phase shift keying (PSK) lie a long time in the future. In equivalent radio terms, the current generation of optical transmitters are akin to the spark gap transmitters and crystal detectors used in the very early days of radio. Compared to optimally efficient (in terms of bandwidth utilization) modulation techniques for single voice channels or single teletype channels, these techniques are orders of magnitude broader.

This does not imply that very broad information bandwidths are not presently available. It does well to caution that a simple and direct comparison between the bandwidths required for narrow channel data on a radio and a fiberoptic link are not necessarily in order. For example, the notation indicating that an optical fiber has a bandwidth/length product of 400 MHz/km does not necessarily mean that a one km link of this fiber could handle $400,000,000/3,000$, or $133,333$ different telephone conversations simultaneously using existing sources and detectors.

Comments in enthusiastic press releases claim that a single fiber can transmit 200 books in a single second. As an exercise, it is perhaps worthwhile to examine this statement in a bit more detail. A typical book such as this one will have perhaps 300 pages of 470 words per page for a total of $141,000$ words. If the average word is made up of five characters and a space to be sent as an 8-bit ASCII code, there would be 6.77×10^6 dot cycles per book. The 200 books per second would come out to be 1.354×10^9 dot cycles per second. While such achievements are not to be dismissed as impossible, they are a far cry from the detectors and sources available today.

The actual information transfer rate achievable in a fiberoptic link is something smaller than this. While this figure is climbing rapidly, it will take a good many years before such performance is seriously challenged. In fibers undergoing current service tests, the goal is a more modest 672 conversations per fiber.

One of the basic limitations on fiberoptic transmission comes from the characteristics of the source. The light-emitting diodes and solid-state lasers used as sources are not very amenable to amplitude modulation. In amplitude modulation, there must be some resting or no-modulation level of signal sent at all times. The information is then impressed by raising the level and lowering the level from the resting level. This is not a particularly efficient process, even in radio. Half of the total power is in the carrier and another quarter is in a superfluous sideband. Single sideband transmission avoids some of these losses, however, as the carrier insertion problem is nearly insuperable.

Frequency and phase shift modulation techniques are also ruled out for reasons of stability. Accordingly, some form of modulation seems to be the most likely candidate. Compared to linear amplitude modulation, pulse modulation does not require any linear relationship between the modulation signal and the output. The pulse is merely turned on or off, which is much easier to do. Secondly, in any analog or amplitude modulation scheme, high levels of modulation in the down direction wind up with little, if any, power output. With no signal being transmitted, a distant receiver will have only a noise output during the dips. By contrast, most pulse modulation schemes have little variation in the power level, and some have none at all. This can also be helpful, because many of the solid-state transmitters must be controlled with regard to temperature. A device which is changing in output level over a wide range is much more difficult to control.

The typical scheme used for encoding the human voice for telephone transmission involves a sampling of the voltage analog of the sound pressure at 8000 times per second. This satisfies the criterion: The least number of samples per wave period to provide meaningful results is two. Figure 3-3 shows a general schematic of the arrangement used in commercial digital telephony. Each sample is applied to a fast analog-to-digital converter.

The A/D converters used in telephony are deliberately arranged to be nonlinear, unlike the A/D devices used for ordinary measurements. This process allows for the *companding* (*com*pression and expan*ding*) of the waveform. The largest and the smallest amplitudes are known to provide less intelligibility and voice recognition than the central range; therefore, the A/D is designed to give more codes and hence more resolution to the central group than to the extremes. There are two different companding laws in use: One is used in the US, and the other, called the CCITT law, is used in most of the rest of the world. The details of this companding are beyond the scope of this text. Companding has the general effect of making the code denser; that is, there is more variation from character to character when companding is used. At the receiving end, a D/A with the same law is used to expand the waveform to the original state.

Pulse Modulation

Two common forms of pulse modulation are shown at the bottom of Fig. 3-3. In the upper waveform, if each bit position or cell is considered, a pulse at the start of the cell represents a 0, whereas

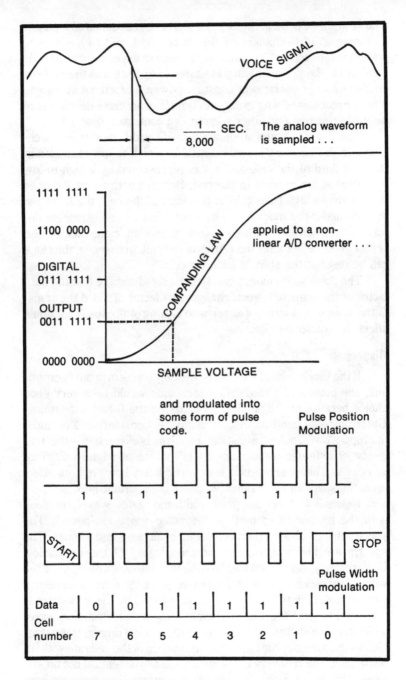

VOICE SIGNAL

$\frac{1}{8,000}$ SEC.   The analog waveform is sampled . . .

DIGITAL OUTPUT

1111 1111
1100 0000
0111 1111
0011 1111
0000 0000

COMPANDING LAW

applied to a non-linear A/D converter . . .

SAMPLE VOLTAGE

and modulated into some form of pulse code.

Pulse Position Modulation

1   1   1   1   1   1   1   1   1

START   STOP

Pulse Width modulation

Data	0	0	1	1	1	1	1	1
Cell number	7	6	5	4	3	2	1	0

Fig. 3-3. Digital telephone techniques.

a pulse at the ¾ point in the cell is a 1. There are certain advantages to having all of the pulses of the same width and to having one transition for each cell, regardless of whether the contents is 0 or 1. First of all, the drive circuitry is easier to operate and the LED or solid-state laser operates at constant power. Unfortunately, timing must be maintained with great precision if an accurate decision is to be made regarding whether a given cell contains a 0 or a 1.

The lower waveform shows a pulse width modulation technique. In this case, a 0 is represented by a high-to-low transition at the first third of the cell, and a 1 is represented by a high-to-low transition at the ⅔ point in the cell. Because both zeros and ones have a low-to-high transition at the start of the cell, the logic can operate under the much less stringent requirement of merely determining whether the high-to-low transition came before the center of the cell or after the center of the cell. In essence, the clock can be reset at the start of each bit cell.

The pulse width modulation has the disadvantage that the duty factor of the transmitter can change by a factor of 2:1 if the transmitter goes from a string of all zeros to a string of all ones. This would affect the transmitter heating.

Dispersion

If the varying duty factor were the only problem in the fiberoptic link, the pulse width modulation technique would be a very good choice because of the simplicity of making the 0 and 1 decisions. Unfortunately, another matter is to be considered. For most fiberoptic waveguides, a distinct problem is *dispersion*—the tendency of different frequencies to propagate at slightly different velocities. This phenomenon is covered later; however, its effect upon the shape of the modulated pulse is important now.

Figure 3-4 shows the pulse width modulated waveform along with the harmonics to n=9 of the pulse repetition period. The spectrum is given for a train consisting of all zeros and all ones. The spectra are identical, except for the phasing of the harmonics. Unfortunately, harmonics well beyond the ninth must be transmitted without appreciable phase shift if anything like the original waveform with ⅓ and ⅔ duty cycle is to be recovered. If the waveform is truncated beyond the fifth harmonic, only a distorted sine wave, which is phase shifted with respect to the original timing, is present. Unless the system propagates the harmonics of the cell rate without phase shift, the recovery of the data becomes difficult, if not impossible. This is not a problem on short systems; however, on long systems, it can make for an unusable system.

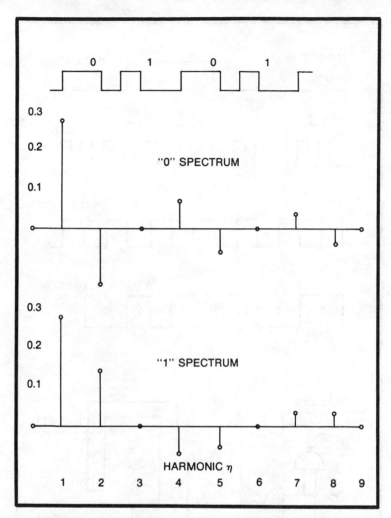

Fig. 3-4. Pulse width modulation.

The problem of waveform distortion in harmonics is not confined to fiberoptic transmission. The digital recording industry has a very similar problem. One of the techniques used to alleviate this problem in high-speed, high-density recording is the technique employed by the Hewlett-Packard Co. in its disk recording systems. This technique is called NRZ/FM recording. Figure 3-5 shows this technique. An initial non-return-to-zero wave train, which is coherent with a data clock and a 2X data clock, is supplied. If the data for a given cell is a zero, the state changes at the end of the cell. If the cell

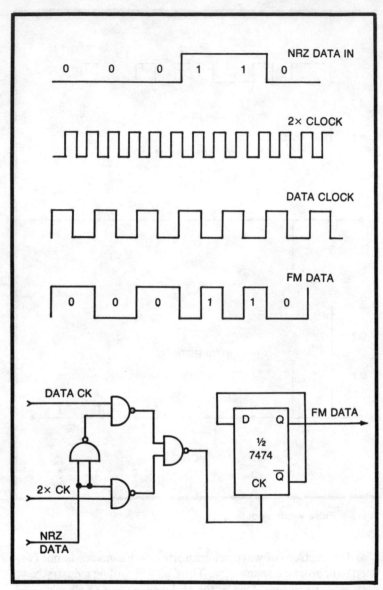

Fig. 3-5. NRZ/FM data modulator. The FM data format operates by providing a level change at the end of each bit cell. The 1's are distinguished from the 0's by the presence of an additional level change in the middle of the data cell. The direction of any level change has no significance; only the presence or absence of a change in the center of the cell does. In effect, the 1's have twice the operating frequencey of the 0's.

data changes at the center of the cell, the figure may be interpreted as a data 1. The circuitry to accomplish this modulation is simplicity itself, as shown on the lower figure.

This technique has several advantages. First, the characteristics of the zeros and the ones are very different. In effect, the ones are operating at nearly twice the frequency of the zeros. The signal can be degraded by phase shifts to nearly sine waves, and the information can be accurately reconstructed.

The second advantage using this technique in fiberoptic transmitters is that the transmitter duty factor is independent of whether the data is all zeros or all ones. It is even independent of the baud rate.

It is easy to see how such a wavetrain can be generated, but it is a little less obvious how one would go about detecting it. Figure 3-5 shows this. FM data from the photodiode or avalanche diode detector are applied to a differentiator and rectifier. If the data are degraded nearly to sine waves, it is only necessary to detect the zero-level or average-level crossing times. This is differentiated and rectified data (DRD is applied to a one-shot with a period equal to ⅔ of a bit cell). If a DRD spike occurrs while the OS is high, the AND gate presets FF-1. After clocking through two flip-flops, the original NRZ data format is restored with a phase shift of one cell. The OS output is also usable as a system clock for subsequent handling of the NRZ data.

One advantage to the NRZ/FM technique that is not immediately apparent is that there are no components in the FM data with a period of less than one-half the cell period. The system is very forgiving of slow rise times on detectors and transmitters and will tolerate very large phase shifts on transmission. For a given link quality, the NRZ/FM modulation technique will generally be capable of operation at something like four or more times the data rate which could have been achieved with the pulse width modulation (PWM) technique.

An interesting approach to creating a constant pulse width makes use of a pulse train from the transmitter, which looks like the DRD train on Fig. 3-6. Like the pulse width modulation technique, this would have the property that the transmitting duty factor would shift by a 2:1 ratio from an all zero to an all one format; however, a constant pulse shape could be used. If the detector were the same speed, though, this technique would be usable only up to about one-quarter to one-third of the rate achievable with the NRZ/FM

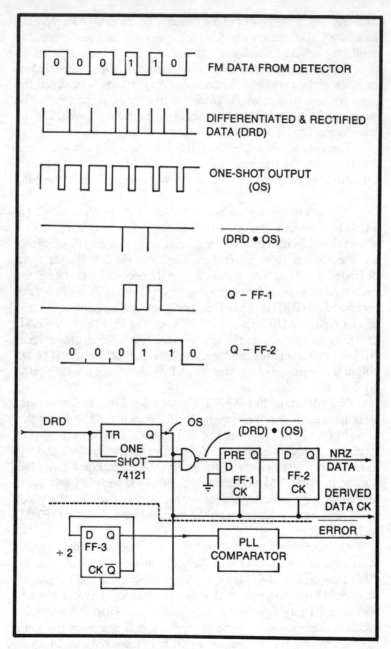

Fig. 3-6. FM/NRZ demodulator.

technique, because the detector would have to detect twice as many pulses.

There are a variety of other modulation techniques for varying pulse trains to impress information upon them. Some have distinct advantages in certain cases. The literature on the subject of modulation techniques and data conserving codes is very extensive. Techniques for digital recording are also of interest.

4

The Nature of Light

This section investigates some of the basic properties of light and the impact of these properties upon fiberoptic devices. Unfortunately, at least from a philosophical point of view, there is still somewhat of a conundrum involved in the study of light. To the physicist, light is composed of quanta of energy, each with a well prescribed wave number. Light is granular in nature. On the other hand, to the radio and communications engineer, light is a smooth electromagnetic wave train which can exhibit interference phenomena, polarization, etc. To the lens designer, light is describable in terms of rays which bounce from surfaces and are sharply bent when they transfer from a medium of one density to another. This picture is an oversimplification of each of the views; however, it does contain the essence of the positions.

While the theoretical physicist may have little difficulty with reconciling the concepts involved in describing the various actions of light, the typical communications engineer does, and often wants to know, what light *really* is. From a pragmatic engineering viewpoint, the communications man is left with the problem that light is really three different things and the correct interpretation is the one which works in solving the problem at hand. When he is working with light-emitting diodes, solid-state lasers, and photodetectors, light is a quantum, a discrete particle with a defined energy. When he is working with a diffraction grating or a single-mode waveguide, light is a wave. When working with impedance matching and diffraction of lenses and reflectors, light is also an electromagnetic wave. When studying the paths within a clad multimode fiber or a graded index fiber, light can be considered to consist of rays, as when describing the action of a lens. Each of these pictures of light has some usefulness to the engineer, and the correct one to use is the one which yields the easiest solution to the problem. Probably the easiest way to put this picture in focus is to give a brief history of the subject. From this you might be able to arrive at a better understanding of the disparities in the various concepts of the nature of light because the various concepts in something like the order in which they were tackled and debated by the scientific community will be presented.

THE SPEED OF LIGHT

Some of the pieces of this puzzle come from rather unexpected places. This account will begin with Galileo Galilei (1564-1642). In 1584, as a 20-year old medical student in the University of Pisa, Galileo noted that the chandelier in the cathedral had a period of swing that required a fixed number of beats of his pulse. Galileo then proceeded to do something that was relatively unheard of in his day. He performed some physical experiments. Soon, he had arrived at a length of pendulum which would permit the very precise timing of the pulse of a patient.

The recourse to experiment is not to be underrated as a leap ahead in science. At the time of Galileo, the established wisdom rested heavily upon the foundations of Euclid (c. 300 B.C.). Euclid's methods as taught in *Elements* relied heavily upon the mental concepts and mental, rational experiments. The straight line which connects any two points and had no dimensions is obviously a mental concept only; it can never actually be constructed physically. The best physical approximation can always be shown to have errors. Thus, the reluctance to accept experimental evidence in those times was not as stubborn and irrational as it might seem today.

Galileo discovered that one of the remarkable properties of the pendulum was that the period of the pendulum had only to do with the length of the pendulum and not with the weight of the bob. Given either a light or a heavy bob, the pendulum would beat at the same rate if the length were held constant. In fairness to the nonexperimentalists, this is also an idealized experiment. A real pendulum does vary its period somewhat with the width of the swing, the mass distribution of the bob, and the mass of the arm; however, these effects are small and were not detectable at the time of Galileo.

The conclusion that the period of the pendulum was a function only of the length led to another significant conclusion which flew in the face of established knowledge. Because the pendulum operates by cycling between gravitational potential energy and kinetic energy (the energy of motion), the conclusion implied that all bodies fall at the same rate. Now this was counter to the wisdom of Aristotle (384-322 B.C.) who had reasoned that heavy bodies fall faster than light bodies. In fairness, this also flew in the face of observation.

Drop a stone and a feather or a leaf at the same time and see which hits the ground first. It is only when the two objects have about the same ratio of mass to air resistance that the apocryphal experiment of dropping a large and a small stone from the tower of Pisa will work with any accuracy.

In 1595 Galileo invented a new technique for polishing lenses which were already widely used as an aid to eyesight for the middle-aged. Galileo also invented a new form of telescope. By 1610 he was selling these telescopes throughout Europe. With these telescopes Galileo was able to discover the four brightest satellites of Jupiter and that Saturn had a peculiar nonround form.

Claudius Ptolemy

By 1600 the earth-centered view of the universe set forth by Claudius Ptolemy (c. 150 A.D.) was beginning to crumble under the weight of its own complexity. Ptolemy had correctly deduced that the moon is closer than the sun or the planets and that the stars are still farther away. He had also prepared a volume catalog of the stars and tables of planetary motion containing 1022 stars. He had discovered the irregularity in the orbit of the moon. In a volume on optics, he had discussed the refraction of light and included a table of the refractive indices of various substances. In geography, Ptolemy had mapped the earth, although he underestimated the size of the Atlantic Ocean and exaggerated the size of Spain and China. As it turns out, these mistakes probably led to the discovery of the New World by Columbus. It has been argued that if Columbus had known the true distance to the Orient, he would have been unable to finance the voyage!

Nicholas Copernicus

The troublesome part about the Ptolemaic theory of the motions of the planets was that it required the planets to move at uneven rates on rather complicated epicyclic paths. The alternate theory proposed in 1543 by a Polish monk and astronomer, Nicholas Copernicus, placed the sun in the center of the universe with the earth revolving about it along with the planets. The moon was proposed to revolve about the earth. This system was much easier to reconcile with increasingly more accurate measurements of planetary positions.

Unfortunately, the Copernican theory posed some rather considerable problems to the Church. First of all it relegated the earth to the position of being just one of a series of planets revolving about the sun. Secondly, it directly contradicted the scriptures, for Joshua had commanded the Sun and the Moon to stand still (Joshua 10:12-14). He had not asked the earth to cease rotating. In addition to all this, the church was involved in the political turmoil of the Reformation; therefore, new positions which questioned traditional authority were generally suspect on political grounds.

Johannes Kepler

In 1619 Johannes Kepler published a tract setting forth his second and third laws. Using the very careful and precise measurements of the Danish astronomer Tycho Brahe, Kepler had earlier demonstrated his first law: Planets revolve in elliptical orbits with the sun at one focus. This law stemmed from the difficulty of trying to fit the measured positions of Mars onto a circular orbit.

When the measured data did not fit the circular scheme, Kepler investigated the ellipse and found that a very close fit was easily obtained. Kepler's second and third laws, however, played more directly into the investigation of the light and gravitation. These are:

☐ The area swept out by the radius vector between the planet and the sun sweeps a constant area in unit time. (When the body is closer to the sun, its velocity is higher.)

☐ The period (time required for one complete circuit) of a planet and the mean distance are related by:

$$\frac{(\text{period } 1)^2}{(\text{period } 2)^2} = \frac{(\text{mean distance } 1)^3}{(\text{mean distance } 2)^3}$$

The third law was significant because it gave a mathematical proportionality for the size of the solar system. The periods of the planets were very well known and cataloged, so the ratios of the orbits of the planets were easily calculated by Kepler. If only one of the orbits in absolute terms could be measured, the size of the solar system and all of the orbits would be known.

From the very earliest times, philosophers had speculated about the velocity of light. From watching a man hammer a stone at a distance of a hundred meters or more, it is relatively obvious that light travels much faster than sound—the hammer is seen to strike the rock long before the impact is heard. However, no one had any way of knowing how much faster light was than sound. As late as 1630, Galileo had attempted to measure the velocity of light by flashing a lantern to a tower at a known distance and having someone there flash a lantern in return when the first flash was seen. Galileo realized that the delays in response completely obscured the travel time of the light and concluded that light must be very fast indeed— perhaps, in fact, the velocity was infinite. The matter would stand there for some years.

Isaac Newton and Christiaan Huygens

The next two major characters to enter upon the scene were Sir Isaac Newton (1642-1727) and Christiaan Huygens (1629-1695). Next to Newton, Huygens was probably the most influential scientist of the day. Based upon the works and findings of Galileo and Kepler, Newton concluded his first investigation and worked out of the laws of gravitation in 1666. These works were laid aside and not published, though. It was not until 1684 that Newton was persuaded to re-examine these findings, which were finally published in 1687 in *Philosophia naturalis principia mathematica (Mathematical Principles of Natural Philosophy)*. This work not only contained the laws of gravitation but also the calculus.

Newton's studies in optics were set forth in Optiks in 1704 in which he presented the results of his experiments in breaking up "white" sunlight into its constituent colors. The study of the spectrum of sunlight would not only lead us to an understanding of the nature of colors, but it would also eventually lead to the development of the quantum theory.

Christiaan Huygens was a Dutch mathematician, physicist, and astronomer. In 1655, Huygens developed a technique for lens grinding which was improved over that of Galileo's. The improvement of the quality of the lenses produced by Huygens' method was so great that he was able for the first time to resolve the rings of Saturn, which lent considerable stature to his reputation. However, another achievement in 1656 was more nearly related to our discussion here. Huygens developed the first quantitively developed theory of Galileo's pendulum and then brilliantly applied this new knowledge to the development of the horologium or pendulum clock. Although clocks had been built with verge and other escapements for more than a hundred years, the Horologium was the first clock which had an accuracy of a few seconds a day. The impact of this upon astronomy was tremendous.

The positions of the stars were largely determined by the time at which the stars passed the meridian. The improvement in timing meant that equivalent improvements in the accuracy of star positions was possible. This also led to improvements in the accuracy of the determination of geographical positions and to an improved knowledge of the size of the earth. The combination of these two achievements led to the induction of Huygens into the British Royal Society in 1663.

This was a period of great geographic exploration. British, French, Dutch, and Spanish ships were traveling the seas of the world in search of spices, treasure, and new territories. However, they were hampered by a lack of precise navigation and positioning. In fact, explorers could not precisely position their new finds upon the map. Latitude can be exactly determined by measuring the height of Polaris or some other known star, or the sun. The accurate determination of longitude, however, requires a precise knowledge of the time. If the time at—say, Greenwich, England—is known, then a measurement of the meridian passage time of a known star or the sun can be used to calculate the longitude from the known rotation rate of the earth, 15 degrees per hour. The problem with this is that after three months at sea, the clocks used for navigation in those days would accumulate large errors that would reflect in an erroneous impression of the longitude.

Gemma Frisius

This method of determination of longitude was originally discussed by a Flemish astronomer, Gemma Frisius, in a book on navigation in 1530. At that time neither the star tables nor the timekeepers were available, though. With the advent of the horologium, the accurate star tables began to accumulate and Huygens set himself to the task of developing a marine horologium. From 1662 to 1670, Huygens constructed and ran sea trials on a variety of specially constructed marine clocks; however the combination of the extreme motion of the ship and the dampness and temperature variation doomed these efforts to failure.

There were two unrelated developments that contributed to this story. If a clock could not be constructed that could be depended upon to keep accurate time during a sea voyage of several months, perhaps there was some technique which could be used to reset an inaccurate clock to precise synchronism with the clock in Greenwich, England or the Hague, where the star tables were based. Today, this can be done with reference to one of the standard radio time signals. How to do it then was the question. The answer proposed was the use of the Gallilean moons of Jupiter. These moons regularly pass into the shadow of Jupiter and wink out in a period of less than a second. In particular, Jupiter I or the innermost satellite is eclipsed every 1.769 days. This very precise event is visible simultaneously over much of the surface of the dark hemisphere of the earth. If a very exact table of these eclipses were available, it would be possible to very precisely synchronize a clock to the standard clock from anywhere on earth. It might take a few days to get the right conditions but it could be done with great accuracy.

Ole Roemer

An astronomer named Ole Roemer (1644-1719) set about to compile such a table. In the course of his measurements, Roemer found a rather curious fact: The eclipses were not as regular as expected and cumulative "late" errors and then "early" errors became noticeable. By 1675 Roemer had accumulated sufficient data to make a significant discovery. Figure 4-1 shows this phenomenon. The curve at the bottom represents the early or late times calculated relative to a straight line from the Roemer data. Roemer postulated that the approximately 1000-second amplitude of the error curve was caused by the fact that the speed of light is not infinite and accounted for the added distance that the light had to traverse going from A to C when the earth was on the opposite side of the sun from Jupiter.

The other unrelated event goes back to the determination by Kepler of the relative size of the orbits of the planets. A number of attempts were made to calculate the distance from the earth to the sun by determining the parallax when the altitude of the sun was measured simultaneously from two points on nearly opposite sides of the earth. This is a very small angle—only 9 seconds of arc—and the measurement must be made in the glare of the sun and in the daytime when the atmosphere is unsteady due to solar heating. Therefore, little success was obtained. At certain times, however, Mars and Earth wind up on the same side of the sun for their closest approach. At these times the Earth-to-Mars distance is only about 80×10^6 km, or about 50 million miles, and thus has about twice the parallax. The measurement is also performed at night when the atmosphere is most steady. One such approach occurred in 1672. Measurements were taken at Paris and French Guiana simultaneously. The measurement produced the result that the parallax of the sun was 8.9 seconds of arc and the mean radius of the Earth's orbit is 149.5×10^6 km, or 93 million miles.

From these experiments, Roemer knew that the distance from point A to point C was 186 million miles; therefore, the 1000 second difference corresponded to a velocity of light of 186,000 miles per second (3×10^8 m/sec). Roemer actually arrived at a velocity of 192 million miles per second due to various inaccuracies; however, this result is still within three percent. It was not until 1850 that Jean B.L. Focault made the first terrestrial measurement of the velocity of light.

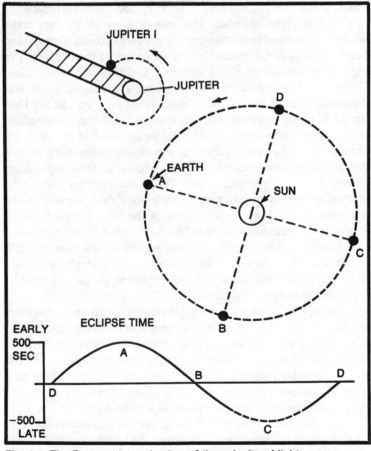

Fig. 4-1. The Roemer determination of the velocity of light.

WAVES OR CORPUSCLES?

Harking back to Newton and Huygens, these contemporary members of the Royal Society had some vastly different views on the subject of the nature of light. Robert Hooke (1635-1703), who developed the spring law governing the elastic behavior of materials and had invented what is known as the Gregorian telescope, had also performed a great number of experiments with light. Hooke formulated the inverse square law (see Equation 3-2) and had postulated a wave theory of light. Newton was unimpressed with these theories for a variety of reasons.

Evangelista Torricelli (1608-1647) had performed an experiment in 1643 in which he filled a glass tube which was sealed at one end with mercury. If the open end of the tube were sealed with a thumb until immersed in a pool of mercury, the column promptly fell to a height of about 30 inches. The volume above the mercury in the sealed tube seemed to contain nothing at all—a vacuum. Toricelli had been a student of Galileo and had replaced him at the University of Pisa. After Galileo's problems with the Holy Office of the Inquisition, Toricelli had little enthusiasm for pursuing a subject which was anathema with the church. Did a vacuum not even contain the Holy Spirit? Toricelli, however, had correctly deduced that the weight of the mercury column was equal to the weight of all of the air in the column above the tube. This idea was quietly communicated to a number of people interested in science who lived a little farther from Rome. A natural question arose: Would the mercury column be shorter atop a mountain where some significant amount of the atmosphere would be below the base of the tube. By 1645, Blaise Pascal had proved that this was in fact the case and had even deduced that the drop in atmospheric pressure was, on the average, one inch per thousand feet. By a linear extrapolation, the atmosphere seemed to be only 30,000 feet high. The space between the planets must be filled with nothing—a vacuum.

To Newton, this was a telling argument. If the space between the planets and the earth were a vacuum, how could light be a matter of waves and waves in what? Furthermore, there was the matter of differential refraction: Different colors of light are refracted by different amounts in passing from one medium to another. This was, after all, the mechanism by which his prism produced a spectrum. Why should waves bend differently depending on color?

The final and strongest argument was the impression that light travels only in straight lines. Why would the light not spread around objects if it were merely waves?

Newton's Theories

For each of these arguments, Newton suggested that light was made of small particles or very tiny corpuscles that traveled at very high velocity. The color was perhaps a manifestation of some physical differences between the particles.

Newton discovered experimentally that when a very slightly convex lens is placed against an optical flat, a series of concentric rainbows surround the point of contact. This phenomenon, which is known today as *Newton's Rings*, is actually a relatively powerful demonstration of optical interference; however, Newton attributed the phenomenon to "fits of transmission and absorption."

Newton also deduced incorrectly that all glasses had the same dispersion of color; therefore, it would be impossible to build a telescope objective which did not suffer from *chromatic aberration*, or have a different focal length for each color. Accordingly, he worked out a telescope design using a reflector for the objective or light-gathering element. In this matter, Newton was also to be proved wrong. Within a few years, Huygens had developed a two-element ocular using identical glass which had almost zero chromatic aberration, although it did have some *barrel distortion*. In 1773, Chester Moor Hall had developed the equations for an achromatic objective with two lenses of different glasses. In 1788, John Dolland produced the first satisfactory achromatic refracting objectives for telescopes.

Christiaan Huygens was an advocate of the wave theory of light, which he had adapted from the work of Hooke. In order to explain the transfer of light across interplanetary space, Huygens invented *ether* (or *aether*) as an invisible weightless fluid which filled all of space. Newton slightly modified his corpuscular theory of light in the publication of *Optiks* in 1704. He allowed his corpuscles to make waves in the ether; however, he pointed out that unless the ether were some kind of a solid, it would only be capable of supporting pressure or sound waves, which are longitudinally polarized. As will be shown, this was a serious blow to the credibility of ether.

In the matter of the waves spreading around obstacles, Huygens had a more successful explanation. The problem was simply this: If light were actually made up of waves, then why did it not simply fill in behind obstacles like ocean waves do behind a piling or a ship? Huygens had adapted the theory set forward by Hooke that each point on a wavefront could be regarded as a new source of disturbance. Figure 4-2 shows the basis of the explanation propounded by Huygens.

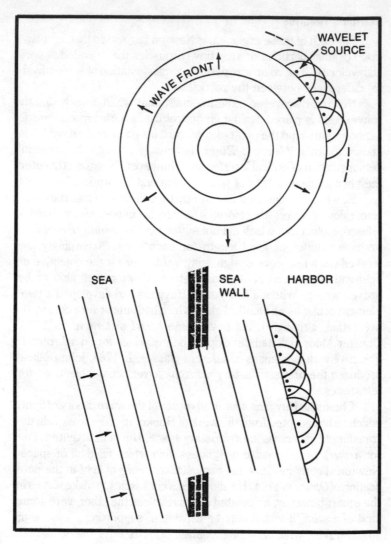

Fig. 4-2. Huygen's wavelets.

For the wave theory to be effective, it would have to explain two familiar but seemingly diverse phenomena. When a pebble is dropped into a pond, the waves propagate outward from the point of impact in a series of concentric circles of equal spacing. On the other hand, when light penetrates a small opening, it seems to propagate in a straight parallel beam or shaft.

The top of the figure shows the pebble-in-the-pond example.

This illustration captures only the crests of the waves that are actually spreading out at a uniform outward velocity. In the upper right quadrant, a section has a number of points on a wave crest that have each been considered to be a new source of disturbance like the original pebble. Each propagates outward at the same velocity. Considered independently, each would produce a circle like the original pebble; however, taken jointly, the only place where they constructively reinforce one another is along the dashed locus of the next wavefront. Everywhere else, they interfere with one another to some extent. Thus, the circular wavefront through the mechanism of the infinite number of wavelets begets another circular wavefront.

Straight Propagation

So much for the circular ripples in the puddle, how about the straight propagation of light beams? The scene in the lower illustration of Fig. 4-2 is a scene that is familiar to those who have spent some time near the sea. A series of long straight-fronted waves from the sea comes in and strikes a sea wall with an opening into a harbor. Only the portions at the harbor mouth get through. If the wavelets are studied, the only place in which they constructively reinforce is along a line that is parallel to the parent wavefront; in other words, a straight-front wave begets another straight-fronted wave. Several other interesting points can be drawn from Fig. 4-2.

First, the development of parallel-fronted waves means that the wavetrain will propagate in a direction normal to the wavefront, because a change in direction would require the wavefronts to be out of parallel. Secondly, spreading occurs only at the edge of the wave. If the harbor opening is large with respect to the distance between waves, the spreading will be very slow and the waves will propagate in a tight bundle or "beam" with relatively calm water on either side. Huygens had noted that a very thin thread or wire viewed against a star (as in the crosshairs of a telescope) did not present a distinct edge but rather showed evidence of light inside the image edges. He took this to mean that some of the wave diffraction was actually taking place. Neither Newton nor Huygens had any way of knowing that light waves are very tiny. Therefore, they did not know how small the wave diffraction would be.

Newton died in 1727 with his corpuscular theory of light largely unchallenged. Perhaps because of his giant contributions to our understanding of gravitation and color, 81 years passed before someone knocked a sizable hole in his theory.

5
Wave Mechanisms

This chapter considers some of the aspects of wave propagation to determine what things they might tell us about the behavior of light. Not electrical waves but rather mechanical waves will be examined first, because the action is easier to visualize. From this point, it is a relatively easy step to adjust our mechanical analog into electrical parameters.

It seems that human beings have studied and pondered waves from the beginning of time. The sight of the rolling sea has certainly entertained artists, poets, philosophers, and lovers in more recent times. If early man studied the sea at length, however, he did not write much about *waves*. The Scriptural account of creation mentions the seas and the creatures of the deep but no waves. Noah could manage to navigate the Flood, and Moses could float amid the bulrushes of the Nile and subsequently see the armies of the Pharaoh drowned beneath the Red Sea with nary a mention of a wave. In fact, the reader must persevere through 590 of the 997 pages of the *Canonical Old Testament* (Revised Standard Edition) to *Psalm 42:7* to find the first mention of a wave. Here King David cries, "Deep calleth unto deep at the noises of thy waterspouts; all thy waves and thy billows are gone over me." This reference (c. 1000 B.C.) antedates *The Iliad* by about 200 years.

This paucity of reference to waves is a bit surprising in view of the large amount of navigation which seems to have gone on in the same period. It is facetious to suggest that it might have stemmed from the tendency to avoid tangling with wave equations, however, a lack of understanding of the processes involved probably did contribute to the situation.

The behavior of waves in the sea is arcane and puzzling. A rough sea is often, but not always, associated with a wind. The wind may or may not blow from the direction from which the waves come. Furthermore, waves do not really do what they seem to be doing. The waves travel in some general direction in the open sea; however, a stick dropped into the water mostly just bobs up and down, describing an approximate ellipse and is generally not transported very much in the direction in which the waves are rapidly traveling.

Leonardo Da Vinci noted that "The impetus is much quicker than the water, for it often happens that the wave flees the place of its creation, while the water does not; like the waves made in a field of grain by the wind, where we see the waves running across the field while the grain remains in its place." Leonardo understood that a wave is not an entity in itself, but rather a condition of a medium. The medium itself is not necessarily transported any net distance, but it conveys the energy in the wave by a transfer of potential to kinetic energy and vice versa. In electromagnetic waves, this particular point will ultimately give us some philosophical difficulties; however, for the time being leave the matter at that.

It is important to note that waves need not consist of a continuous train of disturbances, although this is the most common manifestation. A single wave impulse can also propagate in the same manner. The *seiche* and *tsunami* are examples of single wave phenomena which occur naturally. The seiche is generally caused by a depression in atmospheric pressure, and the tsunami is generally attributed to an earthquake. Our attention will be first directed to such single pulses, because they are easier to follow and observe than a full wave train.

THE LONGITUDINAL WAVE

First, examine the behavior of a longitudinal wave. The longitudinal wave is relatively easy to study and can be observed in an experiment with the simplest of apparatus. The longitudinal or compression-expansion wave is similar to the mechanism which propagates sound through the atmosphere or some other medium. Because sound waves cannot be seen directly, this experiment will be performed with a spring.

The apparatus for the experiment is shown in Fig. 5-1. In Fig. 5-1A, a support and scotch-yoke mechanism is arranged so that an impulse can be applied to the spring through the arm. In this experi-

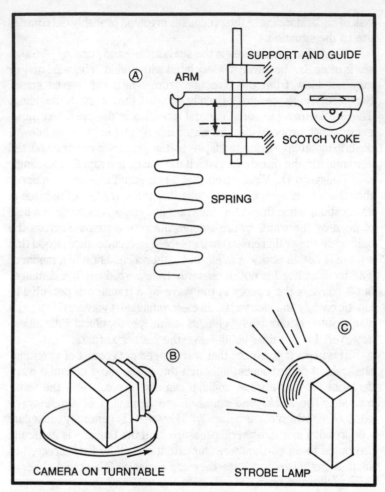

SUPPORT AND GUIDE

(A) ARM

SCOTCH YOKE

SPRING

(B)

(C)

CAMERA ON TURNTABLE

STROBE LAMP

Fig. 5-1. Longitudinal wave demonstration with the apparatus at A, camera at B, and strobe lamp at C.

ment, the impulse will consist of rotating the crank wheel of the scotch yoke through exactly one revolution. This causes the arm to dip down and then return to the starting position and remain there. The spring is a long, limber spring, such as a child's Slinky® toy. It hangs down by gravity.

Figure 5-1B shows a camera mounted on a turntable. Figure 5-1B shows a strobe lamp which repetitively flashes. The turntable and strobe are arranged such that the image of the spring on the film is displaced by the camera rotation just enough so that each flash of the strobe will produce an image of the spring which does not overlap the preceding image. With this mechanism, a picture can be obtained of what the spring is doing in a series of stop action photographs. The result will be a form of motion picture.

The result as presented in Fig. 5-2 was not actually obtained by this mechanism. Instead, a digital computer and plotter were programmed to draw a series of images of the spring at regular time intervals. The computer-generated image, however, is a very close approximation of what the camera would have shown. The camera/strobe analogy gives us a mechanism for visualizing what is happening in the data.

There is another slight difference, though. For the computer simulation, the actual impulse used was a Gauss error function:

$$Y = - e^{- t^2} \qquad \text{Equation 5-1}$$

This function has the advantage that it has only one peak with an amplitude of -1 from $t = -\infty$ to $t = +\infty$. It matches within a few percent a cosine wave from which a single cycle has been cut and displaced one-half unit in the negative direction; thus, it very closely approximates the action of a single turn of the scotch yoke. It begins and ends a trifle more gradually since it is a continuous function. Figure 5-3 shows both this curve and its first and second derivatives. If the value of Y is assumed to be the displacement of a turn in the spring, the first derivative with respect to time is the velocity relative to the center or rest position of the turn. For example, assume that t is measured in seconds. For these computations, assume that the $t=0$ is the peak of the wave and that the impulse begins at $t=-2$ seconds. Actually, a Gauss error function has neither a real beginning or end; however, $t=-2$ and $t=+2$ will suffice for a beginning and end since the displacement is negligible (0.018 units) at this point.

Figure 5-2 begins at $t=-2$ seconds with very little displacement in the top of the spring. By $t=-1$ seconds, the top of the spring has been displaced downward by 0.368 units and a noticeable compressed mass has begun to form ahead of the top turn. From the acceleration curve of Fig. 5-3, the peak acceleration was passed at t equal to 1.2 seconds; however, the peak velocity had not yet been achieved. After $t=-0.7$ seconds, the peak velocity of -0.858 units/sec has been achieved, and the compressed mass is nearly but not quite fully formed. The full formation of the compressed mass has been achieved by -0.5 seconds. At -0.25 seconds, the decel-

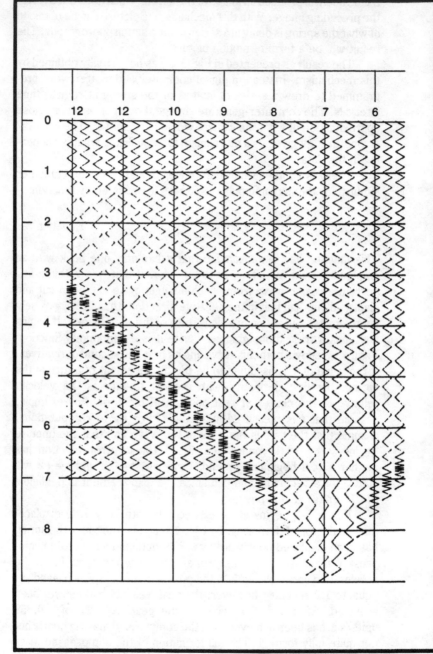

Fig. 5-2. The longitudinal impulse wave.

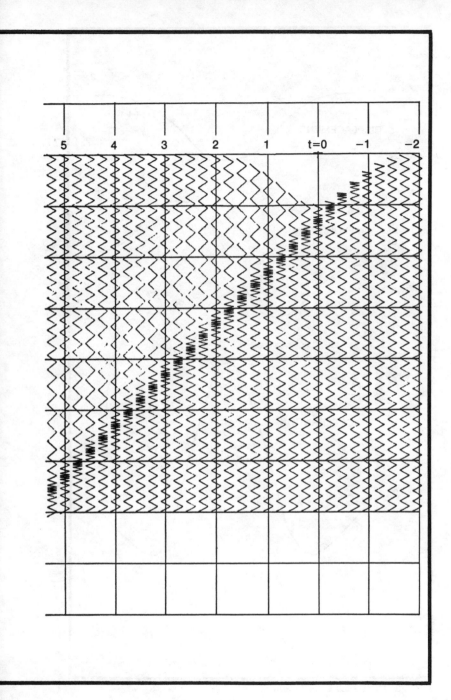

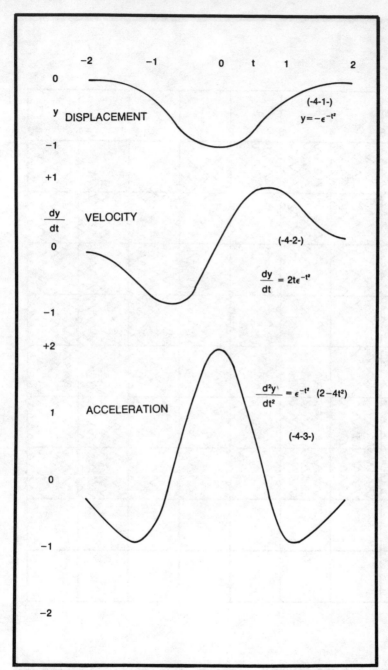

Fig. 5-3. The Gauss error function.

74

eration has caused the compressed mass to detach itself from the top turn.

The maximum acceleration in the upward direction is attained at t=0. Thereafter, acceleration decreases and actually goes negative in the last three quarters of the upward swing. The upper turns reflect this since they are actually in tension and are stretched beyond normal or rest length. The compressed mass has completely detached from the upper section of the spring and is propagating downward at a velocity of one unit per second. At the time t=2 seconds, the impulse is fully developed. Thereafter, the disturbing impulse with tension and compressed mass detaches from the top end of the spring.

If the figure is viewed at a shallow angle more or less along the line of the compressed mass, an optical illusion will be noted in which the entire surface appears as a plane with a swale or trough proceeding from the lower left to the upper right. The apparent cross section of the swale is the Gauss error function which initiated the wave impulse. The dotted curve centered about t=2 seconds is the locus of the tenth turn. It may be seen by examination that the tenth turn is subjected to the same displacement over the same time scale as the initial impulse.

Another noteworthy point at this juncture is that the spring has not behaved as a rigid body. A simple counting of squares will show that the velocity of the compressed mass is one unit per second, whereas the fastest velocity achieved in the impulse was only 0.858 units per second. The wave is fleeing faster than the impulse that formed it. This is the reason that the compressed mass is detached before t=0 at the start of the pulse sequence. This is not an error. The velocity of propagation of the wave down the spring is related to the spring constant and the mass per unit length of the spring and not to the maximum velocity of the starting impulse. In a physical spring, if the maximum velocity of the impulse had been 14 percent higher, the turns in the compressed mass would have crashed into one another to produce a different action. For a sound in air, the equivalent action would be a bow shock, such as the one that forms on a bullet or a supersonic aircraft.

On the other hand, if the initial impulse had attained only half of the propagational velocity, the wave velocity would have been unchanged but the density of the compressed mass would be considerably reduced. It would be harder to discern in the picture, because neither the compression nor the tension would have much contrast with the rest condition of the spring.

If the spring in the example were of infinite length, there would be no more to write about the experiment. After t equals 2 seconds, the impulse is fully detached from the top and would simply propagate in the $-Y$ direction forever. A real spring would have attenuation caused by air friction and hysteresis loss in the steel of the spring wire; therefore, the amplitude of the impulse would decay. In the computer example used to generate the figure, however, no provision for attenuation was included. The nth turn, then, would do exactly the same thing as the first turn. At a time delayed by the travel time from the top, the nth turn would surge forward one unit and settle back to its rest position in a period of approximately 4 seconds.

In the example, the length of the spring was made finite, and as will be discussed shortly, this makes a difference in the behavior of the system as a whole. Examine the bottom end of the spring carefully to note that at $t=5$ seconds, the bottom turn has begun to push beyond the normal rest position. The reason for this is that the compressed mass is beginning to lose spring material to push against. The amount of inertia in the section ahead is decreasing. Therefore, there is less inertial constraint to hold the compressed mass together. At $t=7$ seconds, this phenomenon is fully developed, and only tension is represented in the picture. It is important to note that the extension of the spring at $t=7$ seconds is not one unit but two. The amplitude has grown from one to two units at the open end of the spring.

With the spring fully extended at $t=7$ seconds, it is possible for the bottom turns to accelerate much more rapidly than they can in a continuous section of the spring because they have little mass attached to them to accelerate. By t equals 9 seconds, the spring is nearly back to the rest length and the compressed mass has reformed. The dotted curve of the locus of the 20th turn from t equals 8 to t equals 12 seconds shows us that the rebounding waveform is identical to the initial waveform, except that it is traveling in the opposite direction. At t equals 14 seconds, this rebound would reach the top of the spring.

An examination of Fig. 5-2 shows that the action is the same as if the spring was actually twice as long and an impulse of identical sign (downward) had started simultaneously at both ends of the spring. The impulse started at the top propagates downward, and the impulse started at the bottom propagates upward. At the point seven units below the top, the two impulses pass through one another; thereafter, the downward-going impulse disappears from the bottom of the picture and the upward-going impulse propagates upward in the picture. The amplitude doubles where the impulses cross since they add together. We shall see this same sort of forward-going and backward-going wave phenomenon in some of the following examples.

If you try this experiment with a real Slinky® spring, you will notice several things. First, the extension of the spring under the influence of gravity is not uniform. The top turn is supporting the entire weight and is therefore stretched wide open and each successively lower turn is required to support less weight and is therefore more closed. In fact, if the full spring is used, you cannot lift the main mass of the spring from the floor. If a one-inch high stack of the spring is separated out and the wire marked or notched with a triangular file, the section may usually be broken free by bending at the mark. The lower turns should then be sufficiently stretched to give them a "set" so that the spring hanging in the vertical position has the turns more or less equally spaced. In this condition, the spring will hang about three feet, and it is relatively easy to see the downward impulse propagate and the upward impulse return. The amplitude doubling at the open end is also fairly easily seen.

The *tension/compression* or longitudinally polarized wave is of interest, because this is the only form of waves which can be supported in solids, liquids, or gasses that are uniform and homogeneous. This is to distinguish from gravity waves in water, which exist at the interface of the air and the water. The waves in this discussion propagate entirely in the medium. Some of the other forms of waves which will be discussed will only propagate in solids, which has certain consequences when considering the nature of the ether.

THE TRANSVERSE WAVE

The longitudinal wave is only one of the many types of waves which may propagate through various media. The next example will consider waves in which the impulse is applied not in the direction of travel but rather perpendicular to the direction of travel. This is the general mode by which waves propagate on piano, guitar, violin, and harp strings.

Suppose you take a rope and tie one end to something solid like a post or a tree. If you then stretch the rope out full length with considerable tension and give the near end one shake vertically, you will be able to see the impulse travel at a uniform speed to the fixed end and then reflect back and return to your hand. This is the kind of experiment we will be looking at.

In this experiment, as in the last, the actual trials will be performed upon a computer rather than with a rope, camera, and strobe lamp. The advantage here is even greater than it was for the longitudinal wave because the actual displacements in a real experiment are small and a bit difficult to see on a photograph, much less measure with great accuracy. In this computer-simulated experiment, the disturbing wave can arbitrarily be magnified so that it is visible and accurately measurable.

To have an apparatus to visualize, presume that two of the scotch yoke, support, and arm assemblies have a wire or cable stretched horizontally between them. The two scotch yokes are geared together to make them supply a Gauss error function displacement simultaneously to both ends of the wire. The camera is rearranged to pan upward so that each successively lower image is representative of a later time. After a bit of experimentation, the flash rate of the strobe lamp is adjusted to maximize the three-dimensional effect. Our initial impulse is identical to the one supplied for the longitudinal wave; therefore, the accelerations, displacements, and velocities are identical to those of Fig. 5-3 except for a reversal of sign.

The first experiment is arranged so that both ends of the wire are displaced upward. The flash is started at t equals 0 seconds so it just catches the maximum displacement at the ends of the wire. Figure 5-4 shows the action. Viewed from a shallow angle, the two impulse waves, labeled E_f and E_b, start from the ends of the wire and proceed left to right and right to left, respectively. At $t=5$ seconds, the two waves are fully congruent and their amplitudes add, yielding a peak two units high. The two impulses pass through one another without interference.

In Fig. 5-5, one difference is that the initial impulses are of opposite sign; that is, the impulse starting from the left is downward and the one starting from the right is upward. This figure takes a little more study; however, it may be seen that at $t=5$ seconds—when the pulses are congruent—the wire is straight. The two impulses have completely cancelled one another for a brief moment and pass through one another without other interference. This is an interesting and significant result. Because the point at $X=5$ does not move at all throughout the sequence, it could have been physically attached to a rigid point without changing anything that happened. Return to the rope experiment mentioned earlier, and one property of the two experiments can be verified. With the rope tied to a tree or post, an upward impulse set into the rope initially will be reflected from the fixed end of the rope as a downward impulse. The reflection from a fixed point changes the sign of the impulse.

In electricity, the physical displacement of the rope would actually be translated into a voltage or potential energy function. Because the fixing of the rope to the stationary point does not allow a change in displacement of the rope, the electrical analogue to the fixed point would be a short circuit that will not allow a change in voltage. In a very similar manner, a short circuit on the end of an electrical transmission line or waveguide will produce a reflection with reversed sign, because the boundary condition of the short circuit precludes any voltage from appearing there.

In a similar fashion, the case in Fig. 5-4 corresponds to an open circuit in the transmission line or waveguide. For the rope, this corresponds to a case where the far end of the rope is attached to nothing. With a little practice, you will find it possible to realize this case with the rope as well. The behavior of the rope will be similar to a bullwhip, and the increased amplitude at the open-circuited end will be very noticeable. An initial downward deflection will result in a reflected wave which also has a downward deflection; there is no reversal of sign because the open end can transfer potential energy (displacement or voltage) but cannot transfer kinetic energy (velocity or current).

The behavior in both of these cases is really very much to be expected from the law of conservation of energy. The initial impulse sent down the transmission line, wire, or rope represents a real and finite amount of energy. If the transmission medium is infinitely long, one could expect that this energy could propagate outward forever if the medium is lossless or slowly degrade into heat if the losses are finite. If the medium is finite and terminated in such a way that it cannot absorb energy, however, the energy must be reflected in some manner. In the electrical case the short circuit cannot dissipate energy because $r=0$ and the product i^2r must always be zero. In the open circuit case, r is infinite but i is zero. This results in zero dissipation, also. Therefore, in both of these cases, the law of conservation of energy requires that the energy in the initial impulse must somehow be reflected toward the source because it is not absorbed at the termination.

EQUATIONS FOR THE WAVE

The equations for the waves shown in Figs. 5-4 and 5-5 are described by the functions:

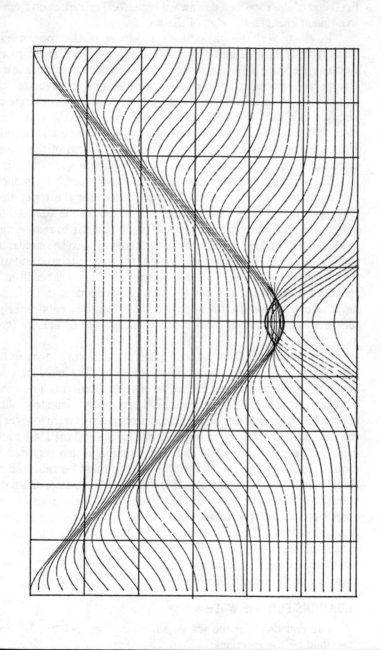

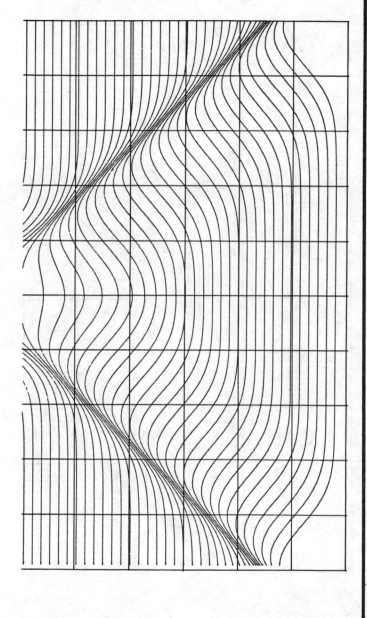

Fig. 5-4. Transverse traveling waves. Ready from the graph grid relative to the flat section of each curve. Time displacement Δt is 0.15 seconds between successive curves.

81

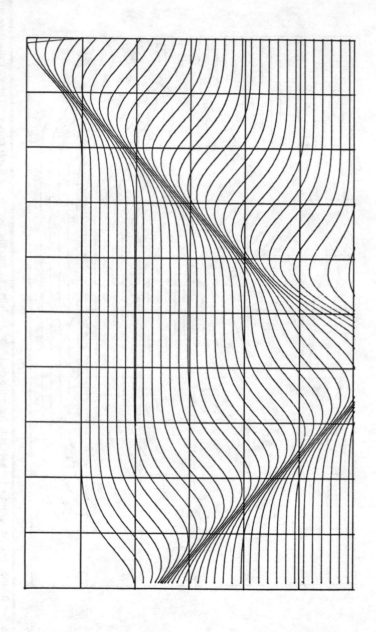

82

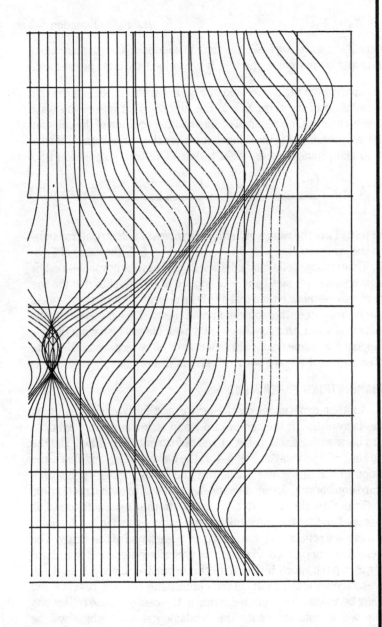

Fig. 5-5. Oppositely signed transverse impulse waves.

$$Y_F = \epsilon^{-\left(t - \frac{x}{V}\right)^2} \qquad \text{for } E_F \qquad \text{Equation 5-2}$$

$$Y_b = \Gamma_b \epsilon^{-\left(t + \left[\frac{x}{V} - 10\right]\right)^2} \qquad \text{for } E_b \qquad \text{Equation 5-3}$$

where ϵ = natural log base, t = time in seconds, X = distance along string in arbitrary units, and V = wave velocity in units per second.

A comparison of the expression of Fig. 5-4 with Fig. 5-1 shows that the inclusion of the term X/V is the function that makes the wave move with time. The reversal of sign and the addition of the constant -10 in Equation 5-3 is the thing that makes the wave from E_b start at X=10 and propagate from right to left. The total expression is:

$$Y = \underbrace{\epsilon^{-\left(t - \frac{x}{V}\right)^2}}_{E_F} + \underbrace{\Gamma_b \epsilon^{-\left(t + \left[\frac{x}{V} - 10\right]\right)^2}}_{E_B} \qquad \text{Equation 5-4}$$

The term Γ_b is the *voltage reflection coefficient* in electrical transmission line work. In Fig. 5-14, $\Gamma_b = +1$. In Fig. 5-5, the value of $\Gamma_b = -1$. This voltage reflection coefficient can have any value—real or imaginary—between +1 and −1. A value of zero implies that the line is perfectly terminated and that all of the energy in the forward-going wave is dissipated, thereby eliminating the backward wave. For the mechanical wave, this condition could be achieved by having the line disappear into some fluid, such as water or oil, which would dissipate the energy in the wave without serious discontinuity.

A PIANO STRING EXPERIMENT

Another experiment can be performed fairly easily if a grand piano is available. If the center of a piano string is given a light but sharp blow with a small mallet, it will deform into something like the shape at t=5 seconds, on Fig. 5-4. If the blow is sharp and short, the tension in the string will accelerate the X=5 (center) point back toward equilibrium. An impulse will flee the spot in each direction and will fly out to the pins at both ends. If the sostenuto pedal is depressed so that the wave toward the hammer end is not damped, the wave will reflect from the fixed pins at each end of the string. The lowest A string on a grand piano will typically yield about five sharp tings at a pitch much higher than the normal string pitch as these waves race back and forth. In the center, the additive reversal will usually be visible. The additive action in the center as the reflections cross will eventually excite the fundamental note which will be present as an aftertone as the tings decay.

VELOCITY OF PROPAGATION

In the previous discussions, the velocity of propagation was assumed to be a constant. Let's consider this assumption in some more quantitative terms. For the time being mechanical waves on wires will still be used because these phenomena are more easily visualized.

String instruments are very old. In about 1000 B.C., the shepherd boy David was summoned to soothe a troubled King Saul because he was "a cunning player upon the harp. . . ." (*1 Samuel 16:16*). Thus, it seems likely that it was implicitly known that the velocity of propagation was a constant for a given string under a given tension. Furthermore, it seems likely that even these early musicians were well aware that the pitch of a vibrating string is:

☐ Directly related to the tension in the string, with a higher tension yielding a higher pitch.

☐ Inversly related to mass per unit length.

☐ Inversely proportional to length.

The human ear, even without training, has a natural mechanism for detecting an octave, which represents a doubling or halving of frequency. The halving of the string length to produce the octave was known from the earliest times. The five-tone scale of the Pythagoreans, the seven-tone scale of India, and the 12-tone scale of China were all derived by linearly dividing the distance along a vibrating string between the open tone and the octave stop.

It is also scarcely credible that a string instrument could be built with a sound box without some appreciation that sound is caused by vibration and the faster vibrations correspond to higher pitches. The effect of the bass notes on a harp are easily seen with the naked eye and felt with the fingers.

With regard to the relationship between tension and mass per unit length, the sailor's trick of twanging a rope under stress to find the amount of reserve strength was also known in antiquity. The sailor knew that too high a pitch indicated that the rope was near its limit of strength. Thus, the fundamental evidence for the wave equations was ancient knowledge at the time of the renaissance. Let's examine some of this evidence in slightly more detail.

Consider some of the forces acting upon a segment of the string in our initial examples. The string shown in Fig. 5-6 is distorted by a wave traveling at a velocity $\vec{V}$. (In this section we have added the arrow over to the symbol to indicate that it is a vector quantity; that is, it has both magnitude and direction.) If a small enough segment of the impulse is considered, it will closely approximate an arc of a circle

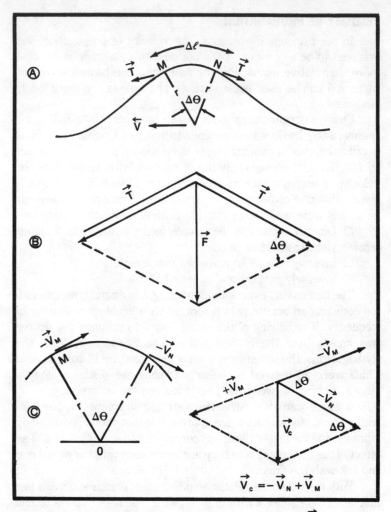

Fig. 5-6. A string distorted by a wave traveling at velocity $\vec{V}$ at A. Vector functions of this effect are shown at B and C.

of radius r. The string tensions $\vec{T}$ act normal to r, and any given point will be whirled through an arc $\Delta\Theta$ as the wave passes at velocity $\vec{V}$. If the arc is sufficiently small:

$$\frac{\vec{F}}{\vec{T}} = \frac{\Delta\ell}{r}$$ Equation 5-5

This statement can be visualized by reference to the vector parallelogram in Fig. 5-6C. The triangles are similar because the included angle is identical. Note that some of the small displacement conditions have been violated in the illustration to make the effect easier to see.

This expression may be resolved into:

$$\vec{F} = \frac{\vec{T} \, \Delta\ell}{r}$$ Equation 5-6

Suppose that you were in a car driving alongside the string at a rate and in the direction of $\vec{V}$. Also, suppose that the car would stay precisely over the center of curvature 0. What you would see is the arc Δl whirling about the center 0. With respect to the moving center 0, this rotation would be just as real as if the string were formed into a complete hoop and rotated about a fixed center. Furthermore, the centrifugal force on the string segment would be just as real. It is this centrifugal force that precisely balances $\vec{F}$ to maintain the wave in equilibrium. At point M, the line is headed outward; at point N, it is headed inward. A given point on the line must therefore have experienced some average acceleration equal to:

$$a = \frac{V_c}{t} \text{ meters / sec}^2$$ Equation 5-7

where a = acceleration, and t = time in seconds to pass between M & N.

From the vector diagram in Fig. 5-6C:

$$\frac{V_c}{V_M} = \frac{\Delta\ell}{r}$$ Equation 5-8

but

$$\Delta\ell = V_M \, t \qquad \text{(meters)}$$ Equation 5-9

Then, substituting from Equations 5-7 and 5-9:

$$V_c = \frac{V_m^2 \, t}{r} = a \, t$$ Equation 5-10

thus:

$$a = \frac{V_m^2}{r} \text{ m/sec}^2 \qquad \text{Equation 5-11}$$

Then in the limit as $\Delta\Theta$ becomes sufficiently small:

$$V_m \cong V$$

For the acceleration of a segment of the line of mass, Δm where Δm is the mass of segment $\Delta\ell$, we obtain:

$$\vec{F} = \Delta m \, a \qquad \text{newtons} \qquad \text{Equation 5-12}$$

and from (-4-13-)

$$\vec{F} = \frac{\Delta m \, V^2}{r} \qquad \text{Equation 5-13}$$

and from (-4-7-)

$$\vec{F} = \vec{T} \frac{\Delta\ell}{r} = \frac{\Delta m \, V^2}{r} \qquad \text{Equation 5-14}$$

thus:

$$V^2 = \vec{T} \frac{\Delta\ell}{\Delta m} \text{ m}^2/\text{sec}^2 \qquad \text{Equation 5-15}$$

The quantity $\Delta m/\Delta\ell$ is simply the reciprocal of the mass per unit length of the string given in kilograms per meter. If this is denoted by m, we obtain:

$$V = \sqrt{\frac{T}{m}} \text{ meters/sec} \qquad \text{Equation 5-16}$$

Note that for small displacements, $\vec{T} \cong T$, where T is the tension in the main body of the string.

Therefore, the velocity of propagation of an arbitrary wave in the string is dependent upon the ratio between the tension and the mass per unit length of the spring. The analysis is subject to a number of limitations that actually boil down to the restriction that the displacement is small. Because the ultimate strength of the string is directly proportional to the mass per unit length for any given material, the trick of twanging a rope to determine how close it is to breaking has a sound theoretical basis. All ropes of the same material loaded to the same percentage of their ultimate strength will show the same propagational velocity.

TRANSLATION TO ELECTRICAL UNITS

We have thus far managed to develop some relations in a mechanical wave system; however, optical waves are electromagnetic. What about the relationships in electromagnetics.

To begin with, mechanical systems have the properties of inertia and acceleration. The electrical duals of these are given in Table 5-1.

Mass m in kilograms is analogous to inductance, and L in henrys and dv/dt in meters per second^2 is analogous to di/dt in amperes per second or d^2Q/dt^2 coulombs per second squared.

This is fine, but it does not give us an analogy for T. T controls the potential energy storage, so it should in some way be related to C in farads. Let's consider the energy storage mechanism in a deflected string. Figure 5-7 shows a string which has been slowly or statically deflected a distance a. The deflection a is so small that it produces a negligible increase in the tension T of the string. For small displacements of the string:

$$\vec{F} = 2\,\vec{T}\sin\Theta \quad \text{newtons} \qquad \text{Equation 5-17}$$

or

$$\vec{F} = 2\,\vec{T}\,\frac{a}{\ell} \qquad \text{Equation 5-18}$$

The incremental energy required to force the string from a to (a + da) is:

$$dw = \vec{F}\ da \quad \text{newton meters or joules} \qquad \text{Equation 5-19}$$

Table 5-1. Certain Mechanical and Electrical Duals.

	Mechanical			Electrical	
Acceleration	$F = m\,\dfrac{dv}{dt}$	Newtons	$E = L\,\dfrac{di}{dt}$	Volts
Energy	$V = \dfrac{mV^2}{2}$	Joules	$W = \dfrac{Li^2}{2}$	Joules

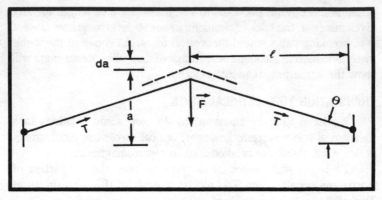

Fig. 5-7. A string that has been slowly or statically deflected a distance a.

Substituting from (-4-22-)

$$dw = 2\,\vec{T}\,\frac{a}{\ell}\,da$$

Equation 5-20

Integrating this expression gives:

$$W = \frac{2\,\vec{T}}{\ell}\int a\;da = \frac{2\,\vec{T}\,a^2}{2\,\ell}$$

Equation 5-21

thus:

$$W = \frac{\vec{T}\,a^2}{\ell}\quad\begin{array}{l}\text{newton meters}\\ \text{or joules}\end{array}$$

Equation 5-22

The expression for the energy stored in a capacitor is given by:

$$W = \frac{Q^2}{2C}\quad\begin{array}{l}\text{joules}\\ \text{(Q in ampere seconds)}\end{array}$$

Equation 5-23

Substituting gives:

$$\frac{\vec{T}\,a^2}{\ell} = \frac{Q^2}{2C}\quad\text{(C in Farads)}$$

Equation 5-24

If we take:

$$\frac{a^2}{\ell} = \frac{Q^2}{2}$$

Equation 5-25

therefore:

$$\vec{T} = \frac{1}{C}$$

Equation 5-26

Substituting this into Equation 5-16 gives the velocity for electrical propagation on the line:

$$V = \sqrt{\frac{T}{m}} = \frac{1}{\sqrt{L\,C}} \text{ meters per second}$$

Equation 5-27

where T = tension in newtons, W = mass/unit length in kg/meter, L = inductance in henrys/meter, and C = capacitance in farads/meter.

For free space, it is possible to separately evaluate the inductance per unit length by using current measurements, and the capacitance per unit length by using electrostatic measurements. The values thus obtained are:

$$u_o = 4\pi \times 10^{-7} \text{ henry/meter}$$

$$e_o = 1/36\pi \times 10^{-9} \text{ farad/meter}$$

Substituting these into Equation 5-27 yields:

$$V = \frac{1}{\sqrt{4\pi \times 10^{-7} \times \dfrac{1}{36\pi} \times 10^{-9}}}$$

$$= 3.00 \times 10^8 \quad \text{meters per second}$$

which is, in fact, the velocity of light in free space as determined by a variety of methods. This analogous approach to the development of expressions for the velocity of light in free space is actually fairly close to the actual historic development. In any event, the analogy has permitted us to make several observations about waves:

☐ Waves propagate by transfer of energy from the potential to the kinetic states and vice versa.

☐ Waves exist in forward and backward traveling pairs which are time functions of $(t - X/V)$ and $(t + ((X/V - K))$, respectively, in order to satisfy the law of conservation of energy.

☐ The velocity of waves is a function of the physical parameters of the medium.

6

Polarization and Wave Interference

From the time of the death of Newton in 1727 until 1800, there was little challenge to the corpuscular theory of light. In 1800, however, Thomas Young, a professor at the Royal Institute in London, showed a theory which purported to explain both the colors to be seen in oil films and soap bubbles and Newton's Rings, which are observed when a lens with a very large radius of curvature is placed against an optical flat. Figure 6-1 shows some of the features of Young's argument. With the two surfaces, there would be some reflection from each surface. When the relationship between the wavelength of the light and the spacing was just right, the two reflections would add because the waves would reinforce one another. On the other hand, there would also be "wrong" spacings where the two reflections would be opposed and tend to cancel. The phenomenon would work best when the thickness was very close to the wavelength of the light. Thus for an extremely thin oil film or soap bubble, certain colors (wavelengths) would add, and certain wavelengths would cancel. Thus the purplish color seen in an oil film could be explained: The red and blue wavelengths were reflecting additively, and the yellow waves which lie between were cancelling. Similarly, the yellow could be explained by a reinforcement of the yellow and a cancellation of the red and blue. Young had correctly deduced that the colors of the spectrum were related to wavelength, but his argument did not sell very well.

One of the big objections to the wave theory of light was that it required a colorless, tasteless, weightless fluid filling interplanetary space to carry the light from the planets and the stars to earth. At that point in time, physics had several weightless fluids. There was *caloric*, which carried heat. There were two kinds of electric fluids: positive and negative. (The unfortunate choice of negative, which is actually an excess of electrons, was made by Benjamin Franklin. He guessed better elsewhere.) The addition of ether to propagate light

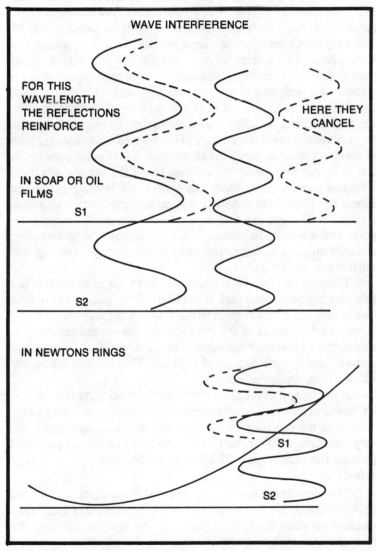

Fig. 6-1. Wave interference.

did not please anyone, because all of these fluids had an ad hoc flavor about them. A number of people were actually striving to rid physics of them.

YOUNG'S WAVE INTERFERENCE DEMONSTRATION

Young persevered and in 1808, he was able to set up an experiment which directly demonstrated wave interference. This experiment is shown in Fig. 6-2. Young could obtain monochromatic (or at least moderately monochromatic light) by heating salts of sodium (yellow), lithium (red), or other salts to incandescence in a burner flame from a spirit lamp. This light was then screened and passed through a single pinhole in a card. It then passed on to a second card with two closely spaced pinholes. This light in turn illuminated a screen at some distance. A curious thing was seen on the target screen. In addition to the two direct beams through the two pinholes, there were a series of illuminated bands flanking both of the direct spots. Figure 5-2B shows this. If each of the pinholes, A and B in the card, were considered to be the source of a new Huygens wavelet, then where the screen was bright represented where the phase shift constant was an integral number of wavelengths, and where the screen was dark represented where the phase shift was an odd number of half wavelengths. The light must be monochromatic; otherwise, the interference patterns overlap and wash out one another.

The purpose of the first, single pinhole card is to assure that the light striking the second card is in phase. If the pinhole in the first card is very small with respect to the distance between the first and second cards, then all of the light striking the second pinhole pair comes from essentially the same Huygens wavelet in the single pinhole—and it must be related in phase. This is called *coherence*, which will be discussed later.

Nowadays, with the powerful monochromatic light of a laser, it is possible to show this experiment in a lighted room. In Young's day, using only the relatively feeble light of the heated salt, however, very little light would sneak through the first pinhole and even less through the second pair. The fringe pattern was therefore very difficult to see.

The experiment had an advantage: It made it possible for the first time to obtain a rough measure of the wavelength of light. If the distance between the bright fringes and the distance between the pinholes could both be measured, the wavelength of the light could be calculated using Equation 6-1 of Fig. 6-2. With an approximately one-meter distance between the pinholes and the screen and a pinhole spacing of d = 0.6 millimeters, red light was found to have fringes about every millimeter on the screen. This yielded a wavelength of 0.6×10^{-6} meters.

The success of this experiment would have been a powerful shove in the direction of the wave theory, but the experiment was

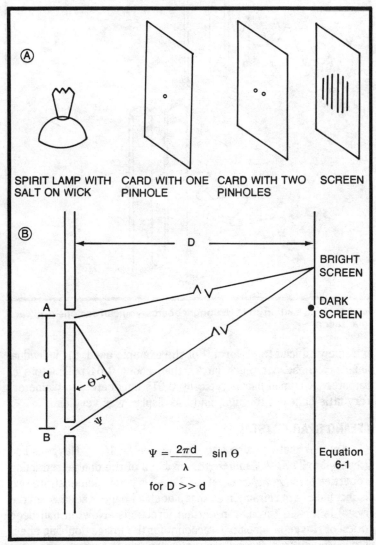

Fig. 6-2. Young's wave interference demonstration setup at A and the result at B.

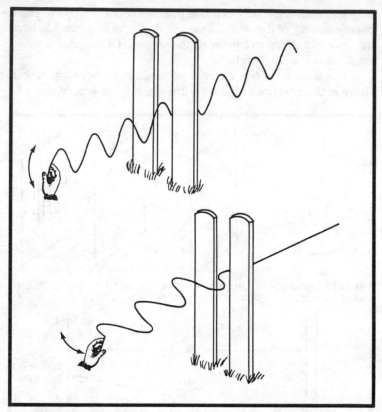

Fig. 6-3. Using a string and two upright boards, you can demonstrate wave polarization.

extremely difficult to perform. For the example used, the individual pinholes can be not much larger than about 0.04 millimeters in diameter. A human hair is typically 0.015 millimeters in diameter. Very little light got through, and the display was very dim.

ICELAND SPAR CRYSTALS

Another matter came into the picture. In 1690, Huygens had discussed in *Trait de Lumiere* the problem of the double refraction properties of *iceland spar* crystals. These crystals demonstrate two distinct indices of refraction so that a double image is seen when the crystal is looked through in certain directions. Newton had been critical of Huygens' inability to account for the image doubling effect with his longitudinal (sound) waves. A break in the situation came in 1808 when Etienne L. Malus reported that light reflected from glass would extinguish when viewed through an iceland spar crystal but could be restored by rotating the crystal about the line of sight. One of the explanations later tendered by Young was that the reflected light might be transversely polarized. Figure 6-3 shows this phenomenon with a rope analogy. If the rope is passed through a picket fence, a wave that is created by shaking the rope up and down will pass freely through the fence, whereas one that is created by shaking from side to side will not pass through the fence.

The iceland spar experiment created more problems for Young. In one attempt to improve the efficiency of his interference experiment, Young had used the iceland spar crystal to give him two pinhole images much brighter than the ones in the second card. The peculiar thing was that these two images would not yield a diffraction or interference pattern.

The answer to this problem was suggested to Young by two different sources. Francis J. Arago noted in a letter to Young in 1816 that light beams will not interfere if they are cross polarized. Consider the two waves of Fig. 6-3. Both waves could coexist without interference since they are orthogonal. The up-and-down oscillation should be entirely independent of the side-to-side oscillation.

The second answer came also in 1816 from Sir David Brewster. Brewster had noted, partially by chance, that a beam of sunlight falling upon a window across the courtyard from his office produced a bright beam which would *not reflect* from a second piece of glass at certain angles. Brewster constructed an apparatus similar to the schematic representation of Fig. 6-4. With it, Brewster began a systematic investigation of the phenomena associated with the reflection of light from glass. Brewster found that the reflection coefficient was very dependent upon the relationship between the plane of polarization and the plane of incidence (B). In fact, he found that a point of zero reflection and 100 percent transmission existed at a well defined angle that was uniquely determined by the indices of refraction of the two media. The symbols $\perp$ and $\parallel$ describe the relationship between the electrical vector and the plane containing the normal to the glass and the angle-of-arrival.

With the Brewster apparatus, which is easily constructed and verified, the existence of transversely polarized light waves was conclusively demonstrated. Young published a complete explanation of the iceland spar problem in an article on chromatics in the 1818 edition of the *Encyclopaedia Britannica*.

In this day and age, when we can perform a verification with a lens from a pair of Polaroid® clip-on sunglasses, it is difficult to

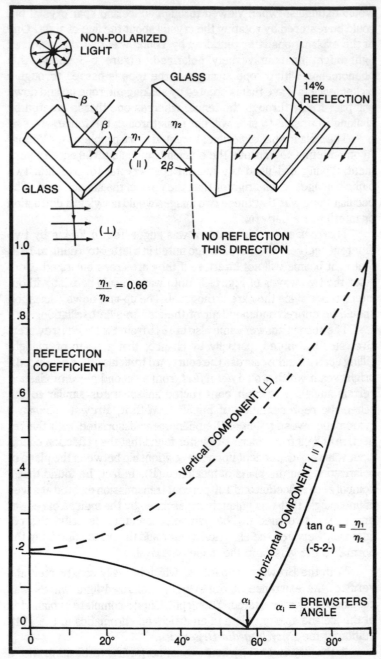

Fig. 6-4. Brewster's apparatus and results.

evaluate the impact of the scientific proof of polarization. There was just no way that one could explain polarization with corpuscles. There were now not just one but two phenomena, wave interference and polarization, which could be explained by wave theory and not by corpuscular theory.

However, if things looked great for the wave theory, the problem of ether became even more severe. As Newton had properly pointed out, it is only the longitudinal wave that can be supported by a gas. The polarization experiments indicated that the light waves were transversely polarized, which meant that ether had to be some form of a solid. Furthermore, ether had to be a solid with some remarkable properties. Young had investigated Hooke's law regarding the elongation of springs and had established the relationship:

$$Y = \frac{\text{Stress}}{\text{Strain}} = \frac{\dfrac{lb}{in^2}}{\dfrac{in}{in}} = lb/in^2$$

This states that the quantity designated as y, which is referred to as Young's modulus, is a measure of the stress on a specimen in lb/in^2, divided by the strain in the specimen in inches of stretch per inch of length.

Relating the quantities to the velocity of sound in a solid, the velocity is given by:

$$V = \sqrt{Y/d}$$

where Y = Youngs modulus for the material, and d = the material density in the same units.

Interestingly, the velocity of sound does not change in quite the way one would expect it to. For example, the value of Young's modulus for steel is much higher than for aluminum; however, steel is also much more dense. Therefore, the velocity of sound in the two materials is about the same, namely about 5100 meters per second compared to 330 m/sec in air. For the velocity of light to be 3×10^8 m/sec, ether would have to have a ratio of Y/d, which is greater by a factor of 3.46×10^9 than that for aluminum or steel. If ether were only a millionth as dense as aluminum, it would still be three thousand times harder to stretch then the finest tool steel. The picture of planets serenely sailing through a material many times stronger than tool steel is not very philosophically satisfying. The wave theory worked, but the ether seemed far fetched. A mechanical transmission of light seemed to be off the track.

This was an era of immense progress with men of science whose interests spanned the length and width of what was known as *natural philosophy.* In 1845, Michael Faraday had discovered that the plane of polarization of light could be twisted measurably by passing the light through glass in the direction of strong magnetic lines. Energize the electromagnet and the plane of polarization twists; turn it off and the glass returns to normal behavior. Because glass is no more magnetic than empty space by all other normal measurements, this *Faraday rotation* implies that light and magnetism are somehow related. Faraday postulated that both light and electricity might be "tensions in the ether." With this development, the stage was set for the creation of the electromagnetic theory of light.

WAVE INTERFERENCE AND DIFFRACTION

The work on the perfection of optical lenses had continued and by about 1810, it had become apparent that there were limitations on the extent of focusing which could be obtained from any lens, no matter how perfectly figured. In 1814 Augustin Fresnel had published a tract dealing with wave interference phenomena. By 1834 G.B. Airy, astronomer royal of England, had published an analysis which stated that the resolving power of any telescope was related to the diameter of the *aperture,* or size of the light-gathering lens or mirror. A mirror or lens which concentrates light to this extent is said to be *diffraction limited.* The following examples shall show how this situation arises.

Figure 6-5A shows much the same situation as Fig. 4-2B, except that the long straight-fronted waves are arriving normal to the aperture. Those in the aperture go through while those striking the screen are reflected or absorbed.

In Fig. 6-5B, for purposes of analysis, the wavefront has been broken into a series of Huygens wavelets. Let's summarize the effect of the wavelets at some distant point P. For the time being, assume that P is so very far distant (compared to the size of the aperture) that the lines from P to any of the Huygens wavelets can be considered parallel. The amplitude of each of the wavelets is designated as E_n and the aperture is divided so that there is a wavelet in the precise center.

Figure 6-5 shows that pairs of wavelets are arrayed symmetrically about the center of the aperture. Therefore, the contribution from the wavelet above the center is delayed at P by the amount Ψ_n. The contribution from the wavelet below the center will arrive early by the amount Ψ_n. Equation 6-1 of Fig. 6-5 gives the exact amount of this phase shift. Equation 6-2 of Fig. 6-5 gives the numerical summation expression for the contribution of all the wavefronts. Before

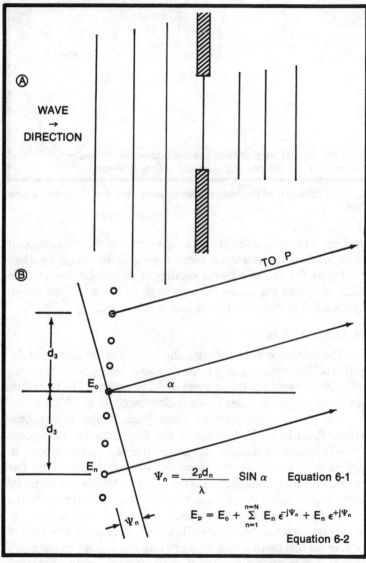

Fig. 6-5. At A, long, straight-fronted waves of a single wavelength impinge normally upon a screen with a rectangular aperture. At B, the wave front is resolved into a series of Huygen's wavelets.

$$\Psi_n = \frac{2_p d_n}{\lambda} \, SIN \, \alpha \qquad \text{Equation 6-1}$$

$$E_p = E_o + \sum_{n=1}^{n=N} E_n \, \epsilon^{-j\Psi_n} + E_n \, \epsilon^{+j\Psi_n} \qquad \text{Equation 6-2}$$

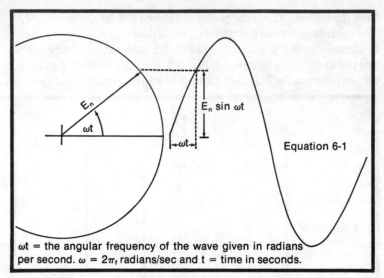

E_n

ωt

$E_n \sin \omega t$

ωt

Equation 6-1

ωt = the angular frequency of the wave given in radians per second. $\omega = 2\pi_f$ radians/sec and t = time in seconds.

Fig. 6-6. The height of the rotation vector above the axis describes a sine wave.

proceeding to develop the physical significance of this expression, it is worthwhile to spend a few minutes reviewing AC vector notation.

Figure 6-6 shows a vector rotating at an angular rate ωt. The height of this vector above the axis is given by the expression of Equation 6-1 of Fig. 6-6. It is in fact a sine wave.

The Complex Plane

The technique of considering the voltage or current in an AC system to be a vector rotating in the complex plane is one of the more powerful techniques in the analysis of AC problems. The technique was introduced by Charles Proteus Steinmetz about 1900.

The phrase, "the complex plane," should also bear a little review. Refer to Fig. 6-7. Suppose that you wanted to have some simple technique for designating the rotation of a vector through an angle of 90 degrees. Further suppose that you decide to call this operator j. If a single application of the operator rotates the vector 90 degrees, two applications rotates it 180 degrees. Then j must be the square root of a minus one. This is an imaginary number, because, of course, no such number exists. However, Steinmetz was fond of pointing out that there is nothing more imaginary about an imaginary number than there is about the distance from Albany to Schenectady.

102

Figure 6-7C shows this point expounded a bit further. If we define the + direction as east, the other definitions follow. Then the distance between Albany and Schenectady can be described as being either 15 miles at an angle of 126.87 degrees (the polar form) or as $-9 + j12$ miles (in complex form).

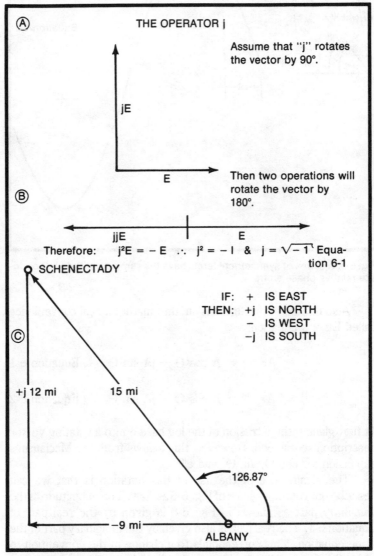

Ⓐ THE OPERATOR j

Assume that "j" rotates the vector by 90°.

jE

E

Then two operations will rotate the vector by 180°.

Ⓑ

jjE   E

Therefore: $j^2E = -E$ ∴ $j^2 = -1$ & $j = \sqrt{-1}$ Equation 6-1

IF: +   IS EAST
THEN: +j IS NORTH
        −   IS WEST
       −j IS SOUTH

Ⓒ

SCHENECTADY

+j 12 mi   15 mi

126.87°

−9 mi

ALBANY

Fig. 6-7. The operator j. Perpendicular at A, opposite at B, and a real-life representation at C.

103

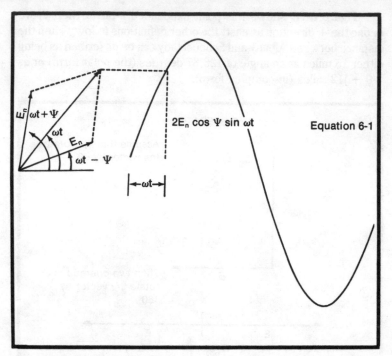

Fig. 6-8. A series of symmetrical terms have the same amplitude but opposite relative phase shifts.

Also related to this figure is another mathematical convenience called Euler's equation:

$$A\epsilon^{j\Theta} = A \cos \Theta + jA \sin \Theta \qquad \text{Equation 6-1}$$

$$A\epsilon^{-j\Theta} = A \cos \Theta - jA \sin \Theta \qquad \text{Equation 6-2}$$

At first glance, the intrusion of the log base e into a rotating vector description seems odd. However, this follows from the Maclaurin's expansion for $\cos \Theta$, $\sin \Theta$, and $e^{j\Theta}$.

The significance of the use of this notation is that we can describe our rotating vector of Fig. 6-6 as $A\epsilon^{j\Theta}$. The magnitude of the *imaginary* part as shown in Fig. 6-6 is given by the real part of Equation 6-1. The real part is also given by the imaginary part of the same equation. This corresponds to a change in the convention to make the figure a little easier to visualize. It does not change the argument to follow.

Multiple Waves

The discussion proceeds to the question of multiple waves that must be summed. From Equation 6-1 of Fig. 6-5, a series of symmetrical terms have the same amplitude but opposite relative phase shifts. This is shown in Fig. 6-8. The two vectors from the + and − sides of the center have positive and negative phase shifts Ψ_n. It is fairly easy to see that the angle between these stays constant with time; after all, Ψ_n is derived only from the geometry of the situation and does not change with time. The whole vector parallelogram does rotate at the rate, ωt, however. This means that the resultant also rotates at rate ωt.

Several conclusions are to be drawn from this. First of all, the addition of a number of sine waves of the same frequency will always result in another sine wave of the same frequency. (Note that this point has not been rigorously proven by the foregoing argument, but the rigorous proof is beyond our requirements here.) The result of the addition will simply produce a new phase and amplitude.

The second conclusion that can be drawn from Fig. 6-8 is that if the amplitude of the two components is equal and the phase is opposite, the phase error of the resultant is zero, $+j$, − or $−j$. Since in our case, the amplitudes of the symmetrical sources are equal, the effect of equal and opposite phase shifts can make the resultant only + or − in phase.

In electrical work, it is sometimes worthwhile to know the absolute instantaneous value of the voltage or current; however, it is only necessary in many cases to know the average or peak or rms value. It is therefore usual to omit the sin ωt term in many computations. The resultant is varying at the rate sin ωt so simply omit that from the calculation. This is even more so in optical work because the frequency is so high that the instantaneous value of the voltage cannot be displayed on any instrument. This means that for any practical purposes, the summation of the contributions for the E's above the center and below the center is given by:

$$E_\Sigma = 2 E_n \cos \Psi_n \qquad \text{Equation 6-3}$$

Figure 6-9 shows, in capsule form, the manner in which these vectors add up. The small vector diagrams assume that the combination of components add up so that at one fixed angle, α, the progression in Ψ_n results. Only + and − values exist in the resultants because the symmetrical E_n's are equal. If $E_{+n} = - E_{-n}$, the results would have all been $+j$ or $−j$.

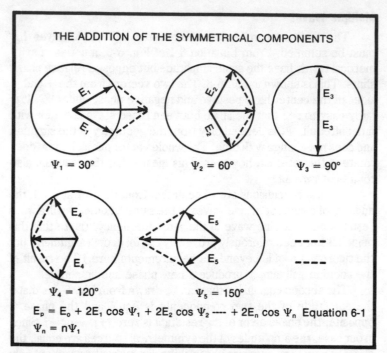

THE ADDITION OF THE SYMMETRICAL COMPONENTS

$\Psi_1 = 30°$

$\Psi_2 = 60°$

$\Psi_3 = 90°$

$\Psi_4 = 120°$

$\Psi_5 = 150°$

$$E_p = E_o + 2E_1 \cos \Psi_1 + 2E_2 \cos \Psi_2 \text{----} + 2E_n \cos \Psi_n \quad \text{Equation 6-1}$$

$$\Psi_n = n\Psi_1$$

Fig. 6-9. The addition of symmetrical components.

Where does the light actually go? To determine this, first sum Equation 6-1 of Fig. 6-9 for as many values of α as desired. It is fairly obvious that for $\alpha = 0$, all of the terms will add up; however, as α departs from zero, the terms with higher values of n will start to cancel first. Then the cancellation begins to spread more rapidly into the lower order terms. The calculation of a radiation pattern by means of a slide rule and a table was obviously a fairly lengthy and tedious process. With a digital computer, it is merely a matter of inserting the correct terms. The machine selects a value of α and then runs the summation for that angle. The angle α is then incremented and the process repeated again and again. Because equation 6-1 of Fig. 6-9 is symmetrical about $\alpha = 0$, it is only necessary to perform the calculation for positive values of α between 0 and 180 degrees.

Figure 6-10 shows one such calculation, performed in terms of Ψ_1. The value of d/λ, which is the spacing of the Huygens wavelets, can be arbitrarily chosen somewhere between 0 and 0.6 wavelengths. If the value is assumed to be 0.5 wavelengths, then the total aperture would be 9.5 wavelengths and $\Psi_1 = 180° \sin \alpha$.

The first zero is located at $\alpha = 7.34$ degrees. The portion inside the first zero is termed the airy disc. The general form of the curve is a sin x/x sort of thing, and the curve becomes a sin x/x curve rigorously if the spacing between the wavelets is taken to be arbitrarily close.

In the original analysis, it was assumed that the aperture was such that all of the E's were equal. This would be a square aperture. However, other aperture types are interesting. For example, in a circular aperture, such as a round lens, the distribution would be collapsed by integration. For such distributions with less illumination on the edges, the beamwidth is somewhat broader and the sidelobes are lower. Table 6-1 gives some of the important functions.

The triangular aperture represents a diagonal cut through a square aperture. Because a of the diagonal aperture is larger than the a of the square (by the square root of two), the beam in the diagonal plane will therefore be broader by the square root of two. But it will have much lower sidelobes, though. Figure 6-11 compares the diffraction patterns obtained from circular and square apertures.

Table 6-1 shows if the aperture starts to be very many wavelengths across, the angular diameter of the airy disc gets very tiny. For example, a round aperture of a = 60 wavelengths would have a first null at 1.14/60 radians or 1.09 degrees off the axis. Yet this aperture would be very tiny at optical wavelengths. For red light, the size would be only $60 \times 0.06 \times 10^{-6}$ meters, or 3.5×10^{-5} meters (1.417 inches) in diameter. This is about the diameter of a thin blond hair.

Babinet's Principle

By a technique called *Babinet's principle*, the shadow cast by an object can be calculated by assuming that the shadow-producing

Table 6-1. Important Illumination Functions.

Illumination	First Zero Angle	First Sidelobe	E_p
Uniform	λ/a rad.	−13.2 db	sin x/x
Circular	1.14 λ/a	−17.1	
Triangular	2 λ/a	−26.4 db	$\left(\sin \frac{x}{2} \middle/ \frac{x}{2}\right)^2$

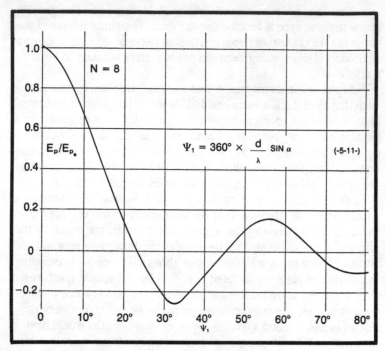

Fig. 6-10. A solution of Equation 6-1 of Fig. 6-9.

obstacle is an aperture and that the aperture is essentially radiating "dark." Thus, a human hair illuminated in red light would cast a shadow with a full width of about one degree. It would have to be viewed from a very short distance to see the diffraction structure.

This explains the earlier question about spreading of light fairly reasonably. For objects the size that people are normally able to view and for apertures the size normally used in optical experiments, the spreading due to the wave nature of light is very tiny. It is only with fine wires or threads or very fine cloth that someone is normally able to view diffraction effects. Newton's assertion that light travels in a straight line is essentially true.

THE DIFFRACTION GRATING

One natural extension of Young's two-hole experiment of Fig. 6-2 would be to construct an apparatus in which there were a large number of equally spaced slots rather than just the two pinholes. Because of the much larger amount of light allowed to pass, the diffraction or interference pattern would be brighter. Consider the properties of Equation 6-1 of Fig. 6-2. Each wavelength will have a different reinforcement point or series of points. This means that our array of slots will have much the same effect as a prism in that it will break up the incident multicolored light into a spectrum or rainbow.

Diffraction gratings can be made either by ruling fine lines upon either speculum metal for a reflective type or upon glass for a transmission type. Less expensive gratings can be purchased which are made by replication. These have a plastic deposited upon a master grating and then stripped off. A common type has the spacing at 14,500 lines per inch. Considering red light at 0.7×10^{-6} meter wavelength and blue at 0.55×10^{-6} meter wavelength and the grating spacing of d = 1.752×10^{-6} meters, Table 6-2 is created from Equation 6-1 of Fig. 6-2. The shorter wavelengths are deviated less.

The various rows are termed *orders of spectra.* In the first order, one wavelength spacing is between adjacent apertures. For the second order, there is two-wavelengths spacing. The line spacing is small enough so that there is not enough room for three wavelengths at the red end between the lines.

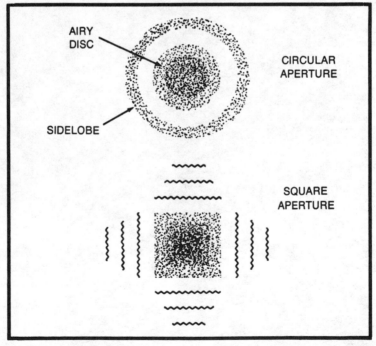

AIRY
DISC

CIRCULAR
APERTURE

SIDELOBE

SQUARE
APERTURE

Fig. 6-11. Diffraction patterns.

Table 6-2. Diffraction Grating Parameters.

Ψ	$\lambda = 0.7 \times 10^{-6}$ Θ	$\lambda = 0.55$ Θ
2π	23.55°	18.30°
4π	53.05°	38.9°
6λ	—	70.38°

On high-quality gratings, it is usual to *blaze* the grating to enhance some particular order, because a refraction spectrometer is seldom used on more than one order. It is not unusual to find in a typical spectrometer that the grating has somewhat larger spacings than used in the example and that the instrument is designed and blazed to use the fifth-order or sixth-order spectrum, because of the superior dispersion available at the higher orders.

7

Wave Impedance

One remaining property of light is best explained electrically, namely the topic of wave impedance. This topic will be discussed next. Then this property will be related to some of the more traditional optical parameters.

LONG-DISTANCE TELEGRAPH

By the early 1850s long-distance telegraph lines had been constructed in a great many places, and it was becoming apparent that telegraph instruments did not behave in exactly the same way when they were separated by many miles of line as they did in the laboratory. The real problem came to a head in 1858 when Cyrus Field had laid the first transatlantic telegraph cable. This initial cable operated successfully for several weeks before failing electrically; however, even if it had been kept in operation, it would have been a financial failure because of the painful slowness with which the telegrams had to be sent.

When the first tests were made, it was found that anything faster than a few words per minute resulted in a hopeless jumbling of the dots and dashes. It took over an hour to send Queen Victoria's greeting to President James Buchanan. Sensitivity was not a particular problem. William Thompson had invented the mirror galvanometer, which would provide a good readable response on steady signals or very slow Morse code. It seemed that the fault lay with the cable itself. It would have to be made to pass faster code if enough telegrams were to be sent to pay for the installation.

Thompson set about to improve the operation of the cable so that the transatlantic telegraph could become a viable financial venture. His success as engineer-in-charge of the revitalized cable company led to honors and fortune. Thompson was knighted Lord William Thompson Kelvin for his brilliant work by Queen Victoria. The 1866 cable financed and directed by Cyrus Field was a technical and business success. News, financial matters, and personal greetings could be flashed across the ocean in seconds with the transatlantic telegraph. The two continents were finally bound together.

Kelvin had been doing work on heat flow and transfer, following ideas suggested by Karl Friedrich Gauss. With this work, he arrived at what have come to be known as *Kelvin's telegraphers equations*. These equations were direct precursors of the work later done by James Clerk Maxwell. The treatment accorded the equations here is simpler than the original. It leans upon some modern concepts not available to Kelvin in his monumental work; however, some effort has been made to retain the flavor of the work.

Figure 7-1 shows physical and electrical representations of a segment of an infinitely long electrical line consisting of two parallel conductors. A battery and switch at one end of the line may be used to start a current flowing down the line. For a small segment Δx, the following voltage drop obtains:

$$\frac{\delta E}{\delta x} = -\left(Ri + L\frac{di}{dt}\right) \text{ volts/meter} \qquad \text{Equation 7-1}$$

This expression simply notes that the voltage drop is proportional to the current times the resistance and the time rate of change of current times inductance.

For the parallel circuit, there is a similar mechanism by which current leaks across the line, both through conductance G and through the charging of capacitor C.

$$\frac{\delta i}{\delta x} = -\left(GE + C\frac{dE}{dt}\right) \text{ amperes/meter} \qquad \text{Equation 7-2}$$

For the time being, the presence of R and G in the equations tends to make the solution a bit more difficult. Therefore, neglect them on the basis that R is very small with respect to L and G is very small with respect to C. This assumption gives us:

$$\frac{\delta E}{\delta x} = -L\frac{di}{dt} \qquad \text{Equation 7-3}$$

and

$$\frac{\delta i}{\delta x} = - C \frac{dE}{dt}$$ Equation 7-4

Next differentiate Equation 7-3 with respect to t and Equation 7-4 with respect to x to obtain:

$$\frac{\delta^2 E}{\delta x \delta t} = - L \frac{\delta^2 i}{\delta t^2}$$ Equation 7-5

and

$$\frac{\delta^2 i}{\delta x^2} = - C \frac{\delta^2 E}{\delta x \delta t}$$ Equation 7-6

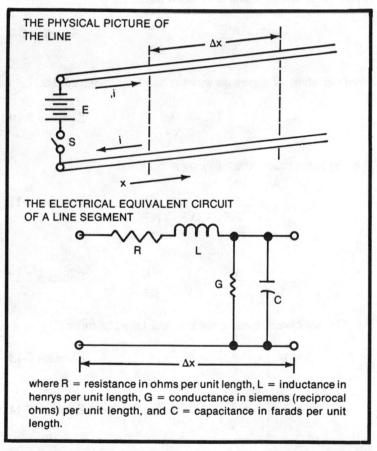

THE PHYSICAL PICTURE OF THE LINE

THE ELECTRICAL EQUIVALENT CIRCUIT OF A LINE SEGMENT

where R = resistance in ohms per unit length, L = inductance in henrys per unit length, G = conductance in siemens (reciprocal ohms) per unit length, and C = capacitance in farads per unit length.

Fig. 7-1. Physical and electrical representations of an infinitely long electrical line consisting of two parallel conductors.

113

Rearranging and substituting gives:

$$\frac{1}{C} \frac{\delta^2 i}{\delta x^2} = L \frac{\delta^2 i}{\delta t^2}$$ Equation 7-7

This can be simplified to:

$$\frac{\delta^2 i}{\delta x^2} = LC \frac{\delta^2 i}{\delta t^2}$$ Equation 7-8

And if Equations 7-3 and 7-4 had been differentiated in the reverse order, the following would have been obtained:

$$\frac{\delta^2 E}{\delta x^2} = LC \frac{\delta^2 E}{\delta t^2}$$ Equation 7-9

From our study of waves on wires it had been deduced that:

$$LC = \frac{1}{V^2}$$ Equation 7-10

Then by substitution, the following is obtained:

$$\frac{\delta^2 i}{\delta x^2} = \frac{1}{V^2} \frac{\delta^2 i}{\delta t^2}$$ Equation 7-11

and

$$\frac{\delta^2 E}{\delta x^2} = \frac{1}{V} \frac{\delta^2 E}{\delta t^2}$$ Equation 7-12

The solution of these equations will take the form:

$$i = i_f \left\{ t - \frac{x}{V} \right\} + i_b \left\{ t + \frac{x}{V} \right\}$$ Equation 7-13

$$E = E_f \left\{ t - \frac{x}{V} \right\} + E_b \left\{ t + \frac{x}{V} \right\}$$ Equation 7-14

Where { } implies only a functional relationship. Note that this is exactly the relationship used for the propagating waves on the wires in Chapter 5. The presence of a minus sign within the brackets will make the waves travel from left to right, and the plus sign will cause them to travel from right to left. Thus, both of the waves demanded by the law of conservation of energy are present.

For compactness, if a_f and a_b are substituted for the quantities within the brackets and you revert to Equations 7-3 and 7-4, the following is obtained:

$$\frac{\delta}{\delta x} (E_f + E_b) = - L \frac{\delta}{\delta t} (i_f + i_b) \qquad \text{Equation 7-15}$$

and

$$\frac{\delta}{\delta x} (i_f + i_b) = - C \frac{\delta}{\delta t} (E_f + E_b) \qquad \text{Equation 7-16}$$

It can be shown that:

$$\frac{\delta E_f}{\delta x} = - \frac{1}{V} \frac{dE_f}{da_f} \qquad \text{Equation 7-17}$$

and similarly:

$$\frac{\delta E_b}{\delta x} = \frac{1}{V} \frac{dE_b}{da_b} \qquad \text{Equation 7-18}$$

A similar analysis obtains for i_f and i_b. Substituting into Equations 7-15 and 7-16, the following is obtained:

$$\frac{1}{V} \left(- \frac{dE_f}{da_f} + \frac{dE_b}{da_b} \right) = - L \left(\frac{di_f}{da_f} + \frac{di_b}{da_b} \right) \qquad \text{Equation 7-19}$$

$$\frac{1}{V} \left(- \frac{di_f}{da_f} + \frac{di_b}{da_b} \right) = - C \left(\frac{dE_f}{da_f} + \frac{dE_b}{da_b} \right) \qquad \text{Equation 7-20}$$

This may be transposed into:

$$\frac{di_f}{da_f} + \frac{di_b}{da_b} = \frac{1}{LV} \left(\frac{dE_f}{da_f} - \frac{dE_b}{da_b} \right) \qquad \text{Equation 7-21}$$

and

$$\frac{di_f}{da_f} - \frac{di_b}{da_b} = CV\left(\frac{dE_f}{da_f} + \frac{dE_b}{da_b}\right)$$ Equation 7-22

Then adding and subtracting yields:

$$\frac{di_f}{da_f} = \frac{1}{LV}\left(\frac{dE_f}{da_f}\right)$$ Equation 7-23

and

$$\frac{di_b}{da_b} = -CV\left(\frac{dE_b}{da_b}\right)$$ Equation 7-24

However, we had noted earlier that:

$$V = \frac{1}{\sqrt{LC}}$$ Equation 7-25

Therefore:

$$\frac{1}{LV} = \frac{\sqrt{LC}}{L} = \frac{1}{\sqrt{\frac{L}{C}}}$$ Equation 7-26

and

$$CV = C\frac{1}{\sqrt{LC}} = \frac{1}{\sqrt{\frac{L}{C}}}$$ Equation 7-27

Returning to Equations 7-23 and 7-24 and integrating and substituting, the following is obtained:

$$i_f = \frac{E_f}{\sqrt{\frac{L}{C}}} + K_f$$ Equation 7-28

116

and

$$i_b = \frac{E_b}{\sqrt{\dfrac{L}{C}}} + K_b \qquad \text{Equation 7-29}$$

The terms K_f and K_b are constants of integration pertaining to DC terms and may be neglected. However, the remainder of the result is of great interest to us because it determines the ratio of the voltages and currents on the line. It must therefore have the dimension of ohms. It is a bit surprising at first to think that:

$$\sqrt{\frac{\text{Henrys / meter}}{\text{Farads / meter}}} = \text{ohms} \qquad \text{Equation 7-30}$$

However, the ratio of volts to amperes is relatively obvious, so the value given by:

$$Z_o = \sqrt{\frac{L}{C}} \text{ ohms}$$

is referred to as the *characteristic impedance of the transmission medium.*

Viewed on a philosophical level, it seems logical that the ratio between the voltage input and the current input on a transmission line must be established by something local. Suppose that the velocity of propagation on the transatlantic cable was 2×10^8 meters per second and the length of the cable is 3000 miles, or 4.8×10^6 meters. It would take 24 milliseconds for the signal to travel from one end to the other, making it fairly easy for the telegrapher to close and open the key before the signal travels from the sending end to the receiving end. Something must have established the ratio between the voltage and current during transit.

REFLECTIONS

Consider the situation of a given wave launched upon the transatlantic cable. At the outset, the voltage/current relationship is determined by the characteristic impedance of the cable. If the cable were infinitely long and there were no losses, the launched wave continues forever. Similarly, if the far end of the cable is terminated in a resistor with a value equal to Z_o, the relationship between the voltage and current is not required to change for the outgoing wave. In this condition, the sending end cannot determine whether the cable is infinite in length or merely properly terminated.

On the other hand, suppose that the far end of the cable is terminated in an open circuit. When the outgoing wave reaches the far end, the voltage/current ratio must be drastically altered; in fact, the current must go to zero. This can be accomplished by a backward wave of current such that $i_b = -i_f$. During the transit time from the far end to the sending end, the voltages of these two waves add. In the case of a simple step wave, such as turning on a battery, the voltage on the line *doubles*.

When the far end of the cable is short-circuited, the opposite situation obtains. The voltage must suddenly fall to zero. This effect is obtained by a backward wave with $E_b = -E_f$, and the current doubles. This is similar to the effect seen in the spring of Fig. 5-2. In the original transatlantic cable, Kelvin noted that the impedance of the detecting instrument must be made equal to the Z_0 of the cable itself. When this was not the case, a severe reflection was sent back. When the reflection arrived at the sending end, it would see either an open circuit or a short circuit, depending upon whether the key was open or closed. If rapid transmission was attempted, the cable became filled with reflection pulses racing back and forth. Consequently, the transmission became jumbled and confused.

The relationship between the forward and backward waves is given by:

$$\frac{E_b}{E_f} = \frac{Z_L - Z_0}{Z_L + Z_0} = \Gamma \qquad \text{Equation 7-31}$$

where Z_l is the load impedance.

Z_l may be either pure resistive or it may be a complex impedance. And it may take on any value from zero to infinity. The voltage reflection coefficient, on the other hand, might take on any real or complex value with any sign, but the magnitude is *always* less than 1. For $Z_l = Z_0$, the voltage reflection coefficient is zero.

In fiberoptic links, particularly low-loss links, the importance of matching the terminating impedance stems mainly from the multiple echo effect, as it did with the transatlantic cable. Each impedance discontinuity gives rise to its own echo, and a high data rate system can wind up by being plagued with a number of stray reflection pulses rattling around in the system. For particular spacings between the discontinuities and for particular pulse spacings, these echoes can build up to substantial amplitudes so that a considerable "trash" background can exist. In general, if the spacing between the discontinuities is such that the wave will attenuate to $1/e$, or $-8.69\,\text{dB}$, the buildup cannot occur; however, for the lowest loss cables, this represents a substantial length of cable and a fairly healthy ring time. For example, on a cable with an apparent (electrical) length of 2 km, a second-time echo would have to travel 4 km. At a baud rate of only 75,000, this could represent the echo arriving at the same time as a new character bit.

In free space, the impedance of the medium is given by the same relationship, namely:

$$Z_0 = \sqrt{\frac{L}{C}} = \sqrt{\frac{\mu_0}{\epsilon_0}} \text{ ohms} \qquad \text{Equation 7-32}$$

Where $\mu_0 = 4\pi \times 10^{-7}$ henrys per meter, and $\epsilon_0 = \dfrac{1}{36\pi} \times 10^{-9}$ farads per meter.

Solving Equation 7-32 gives a value of 377 ohms. This is the relationship between the electric and magnetic components of a plane wave propagating freely in space. For optical waves in other media, the impedance will be modified by the relative dielectric constant of the medium. In general, optically transparent media have no measurable permeability or magnetism higher than free space. Certain glasses do exhibit the Faraday rotation noted earlier, though.

WAVEGUIDES

The use of hollow metal pipes to guide waves is relatively new, dating back mainly to the 1930s. The use of dielectric waveguides is even newer. There is really no counterpart in optical work for transmission lines and coaxial cables because of the extremely small sizes involved. A coaxial cable will only function well when the inside diameter of the cable is less than half a wavelength. A two-wire transmission line must have a spacing that is considerably smaller still—down to 0.1 or 0.05 wavelengths. In fact, there is really no way of generating optical waves that corresponds with the direct electrical generation of radio waves and microwaves by tubes and transistors. For each of these devices, the circuit elements usually need to be a small fraction of a wavelength. This would make the elements impossibly small at optical frequencies. Even devices like the klystron tube and the traveling-wave tube must have structures that are small with respect to wavelength.

Optical Fibers

The optical fibers used in fiberoptic transmission are a counterpart of the waveguide. It is therefore useful to examine some of the principles of waveguides to discover some of the properties. We shall begin with the hollow metal pipe waveguide (although this does not have a practical application at optical wavelengths) because this is the easiest type to show. The discussion shall then extend to cover the single-mode fibers.

To begin with, examine Fig. 7-2. We have resorted again to our waves-on-wires simulation for a visible demonstration of wave interference. Proceeding from the upper left hand corner of Fig. 7-2, a train of sine waves marches across the figure from left to right. About halfway down the page, the wavetrain collides with the obstacle at $x=8$. A reflected wave with $E_b = -E_f$ begins to march backward across the page from right to left. The amplitude of the waves is doubled in places, and a peculiar thing is happening. A careful study of the individual lines will show that the points at $x=7, 6, 5, \ldots 1$ are perfectly stationary, whereas the points in between are oscillating at twice the amplitude of the incoming waves. The pattern of the loops and nodes is repeated every one-half wavelength. Viewing the paper at a slanting angle along the two sets of wavefronts will give a better view of this with the quasi-three-dimensional effect noted earlier. The section in the lower right quadrant of the figure is called a *standing-wave pattern*.

To understand more about this phenomenon, use a bit of arithmetic again. The equations describing the steady state of the waves are:

$$Ex = E_f \, \epsilon^{jw\left(t - \frac{x}{V}\right)} + E_b \, \epsilon^{jw\left(t + \frac{x}{V}\right)} \qquad \text{Equation 7-33}$$

$$i_x = \frac{E_f}{Z_o} \, \epsilon^{jw\left(t - \frac{x}{V}\right)} - \frac{E_b}{Z_o} \epsilon^{jw\left(t + \frac{x}{V}\right)} \qquad \text{Equation 7-34}$$

For a convenient scale on Fig. 7-2, f has been set equal to ½; therefore, the waves are two units long.

Figure 7-3 shows the relationship between the voltage vectors for the forward and backward waves. We can see that the entire ensemble rotates at the angular rate of ω. However, the *difference* between the two vectors is the factor that determines the way in which they add. In Fig. 7-3, this difference does not vary with time but does vary with position x. Furthermore, the difference in sign between the x/V terms means that the difference increases twice as fast as the distance in wavelengths. Therefore, the standing-wave pattern has a period of one-half wavelength, or a single unit in Fig. 7-2.

Note that the computer program which generated Fig. 7-2 was adjusted one-quarter wave from the given equations for reasons of appearance.

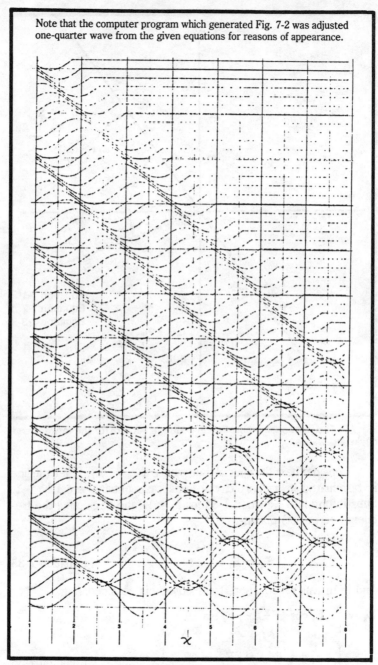

Fig. 7-2. Wave interference.

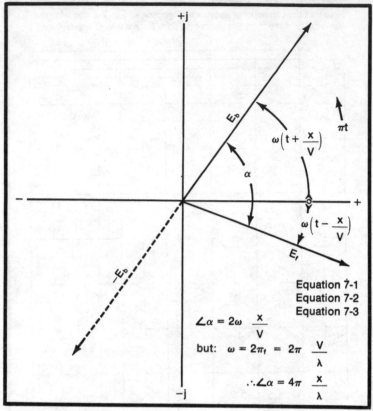

Fig. 7-3. The relationship between the voltage vectors for the forward and backward waves.

Consider next the special case where $E_b = -E_f$. When the line is terminated in a short circuit, this case obtains and $\Gamma = -1$. Substituting into Equation 7-33 for $F = \frac{1}{2}$ and neglecting the frequency term (ωt) obtains:

$$E_x = E_f \, \epsilon^{-j \frac{\pi x}{\lambda}} - E_f \, {}^{+j \frac{\pi x}{\lambda}} \qquad \text{Equation 7-35}$$

And from Euler's equation, the following is obtained:

$$E_x = \left[\; E_f \cos \frac{\pi x}{\lambda} \; -j \, E_f \sin \frac{\pi x}{\lambda} \; \right] - $$
$$\left[\; E_f \cos \frac{\pi x}{\lambda} \; +j \, E_f \sin \frac{\pi x}{\lambda} \; \right] \qquad \text{Equation 7-36}$$

Noting that the cosine terms cancel and sum:

$$E_x = -2jE_f\sin\frac{\pi x}{\lambda}$$ Equation 7-37

Now note that with $E_x = 0$ for $x = \frac{1}{4}$, $\frac{3}{4}$, $\frac{5}{4}$, ... odd $\infty/4$ wavelengths and $E_x = \pm j\,2E_f$ for $x = 0, 0.5, 1, 1.5, \ldots \infty/2$ wavelengths. All of the locations are varying at angular frequency ωt; however, the amplitude is permanently zero at certain locations.

Returning to Equation 7-35, the variation of the current can also be calculated with a location on the line:

$$i_x = \frac{E_f}{Z_o}\epsilon^{-j\frac{\pi x}{\lambda}} + \frac{E_f}{Z_o}\epsilon^{+j\frac{\pi x}{\lambda}}$$ Equation 7-38

In this case, the plus sign of the second term means that with the application of Euler's equation, the imaginary terms cancel and the cosine terms add.

$$i_x = 2\frac{E_f}{Z_o}\cos\frac{\pi x}{\lambda}$$ Equation 7-39

Note that this equation has zeros everywhere that Equation 7-37 has maxima. Also, it has maxima where Equation 7-37 has zeros.

Consider the significance of this for a moment. At a location where the voltage is zero and the current is finite, such as $x=\frac{1}{4}$ wavelengths, the impedance of the line itself must be zero because the line is behaving like a short circuit.

At the locations on the line where the current is zero and the voltage is finite, the line is behaving like an open circuit or an infinite impedance. These points would be located at $x=0, 0.5, 1, \ldots \infty/2$ wavelengths. The apparent impedance on the line cycles between an open circuit and a short circuit every quarter wavelength and back again in another quarter wavelength with the cycle repeating itself every half wavelength. Viewing any segment from the standpoint of impedance, it is impossible to determine from the sending end whether the far end is terminated in an open circuit or a short circuit. Only when the exact length of the line is known can one determine the difference. The impedance patterns are identical except for a quarter-wave displacement.

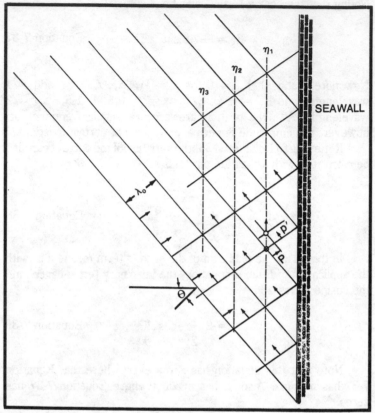

Fig. 7-4. The seawall waveguide and analogy.

Note that the computer program which generated Fig. 7-2 was adjusted one-quarter wave from the given equations for reasons of appearance.

Standing Waves

Standing-wave patterns can frequently be seen in the surface of coffee in a cup aboard an airplane or other vehicle. They can also sometimes be observed in the radio antenna of an automobile if there is appreciable vibration in an idling engine. It sometimes takes a bit of watching, but you'll notice that the portions of the medium that are oscillating violently are spaced between portions that are not moving at all.

For a more general case, consider Fig. 7-4. A series of long, straight waves is impinging upon a straight seawall at some angle Θ. The waves rebound at angle Θ as well. The crests of the waves

124

always cross along the dotted loci parallel to the seawall. For example, note the point designated P. At some time later, the same wavefront has moved toward the wall; however, the rebounding wave has moved away from the wall at the same rate. The new intersection is at P'.

Note that the conditions along the locus P, P' are the same as the conditions along the seawall itself. A second seawall could even be erected along the locus. The waves would still continue to propagate along the channel exactly as they do now, rattling back and forth between the sides. This would be a waveguide.

A second wall inserted along locus n_1 would be termed a *dominant mode* or *first mode* waveguide. The wall at n_2 would be a second mode guide, the wall at n_3 would be a third mode guide, and so on. Note that from our standing-wave expression, equal amplitudes of forward and reverse waves gave points where the impedance was actually zero. It would be possible to insert a perfectly conducting wall along these loci without disturbing the standing-wave pattern between the two discontinuities.

If angle Θ were zero, the spacing to n_1 would be precisely one-half wavelength, as noted in the earlier discussion on standing waves. In Fig. 7-5, points P and P' have been enlarged slightly. The original intersection at P was made up by the intersection of elements C and D. Each of these points travel at right angles to the wavefront; therefore, they flee from P in the directions shown at the bottom of Fig. 7-5. The intersection at P' will not be made of these two points, but rather of points A and B, which shall eventually collide at P'. Because of the inclination of Θ, each of these points has a velocity component toward or away from the seawall and another component parallel to the seawall, as shown in the resolving triangles. In time t, points A and B shall move a distance Vt sin Θ parallel to the seawall. This is termed the *group velocity* of the wave. It corresponds to the *course made good* in sailing. If a pulse consisting of a small packet of sine waves were launched in the waveguide, this would be the velocity at which the pulse would travel along the guide.

On the other hand, consider the action between P and P'. If t is arbitrarily assigned equal to one cycle, Vt is then equal to one wavelength. In this case, if the wavelength were measured along the path P-P', the distance between the wave crests is much larger than the actual wavelength and is increased by the reciprocal of sin Θ. In air-filled metal pipe waveguides, this presents an unusual aspect because the phase velocity is always *greater* than the velocity of light in free space and the apparent wavelength in the waveguide is always

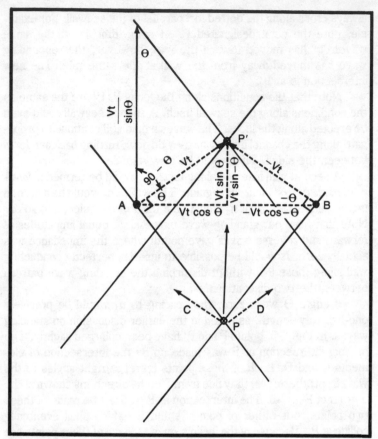

Fig. 7-5. Phase and group velocity.

greater than the wavelength in free space. As Θ goes to zero, in fact, the group velocity goes to zero and the phase velocity goes to infinity!

Obviously, if the waves were approaching perfectly normal to the seawall, all points along the seawall would be in phase, thereby giving an apparent phase velocity of infinity. The group velocity would be zero, implying no net transport of wave energy along the seawall.

For a metal pipe waveguide, if we presume that the space between n_1 and the seawall or the width of our pipe, is fixed, then the largest wave that can be made to propagate at all in the pipe is of a size such that the wavelength is just twice the width of the guide. This is referred to as the *cutoff wavelength* of the waveguide. The group and phase velocity are related to this cutoff wavelength as follows:

$$V_9 = \sqrt{1 - (\lambda_0/\lambda_c)^2} \qquad \text{Equation 7-40}$$

$$\lambda_\phi = \frac{\lambda_0}{\sqrt{1 - (\lambda_0/\lambda_c)^2}} \qquad \text{Equation 7-41}$$

where V_9 = group velocity, V_0 = velocity in free space (or medium), λ_0 = free space wavelength in free space, λ_c = cutoff wavelength, and λ_ϕ = phase wavelength.

These follow from the calculation of the angle Θ and the Pythagorean theorem. The phase wavelength is given rather than the phase velocity, for this is the quantity usually required.

The hollow metal pipe waveguide is generally not useful at optical frequencies. A hollow metal pipe waveguide is usually operated with wavelengths on the order of three-quarters of the cutoff wavelength because the attenuation is usually lowest at this point. The attenuation rises for shorter and longer wavelengths. Because of skin effect, the attenuation per wavelength for this type of guide tends to be essentially independent of frequency. It runs about 1 dB per 300 wavelengths. At optical wavelengths, there are something on the order of 1.7×10^9 wavelengths per kilometer. Therefore, the attenuation would be on the order of 5.6×10^9 dB/km, or more than 5000 dB/meter. Some of the higher-order metal pipe modes have much less loss; however, a mechanism for exciting these has not been developed. In addition, the requirement that a surface finish must have disturbances that are small with respect to wavelength seems to place these devices out of reach of existing technology.

On the other hand, the techniques for drawing fibers from optical glass and plastic are quite adequate for producing fibers capable of supporting a single propagation mode. Waves can be guided by fibers with a diameter as small as a one-quarter wavelength. For such small fibers, the power is mostly outside the fiber, and the phase velocity is determined mostly by the outside medium. As the diameter of the fiber begins to approach a wavelength, most of the power is confined within the fiber. The propagational velocity approaches the propagational velocity in an *unbounded* medium of the same material.

These fibers could be completely homogeneous and still provide the wave-guiding action. The electromagnetic fields of the guided waves, however, would not be completely contained within the fiber and media outside the fiber but adjacent to it would have a tendency to disturb the propagation and introduce both reflections and attenuation.

In general, the fibers used in single-mode applications are not constant dielectric cross sections. Instead, their dielectric changes on the basis of radial distance from the axis of the fiber. We shall be discussing these shortly.

The concepts developed in this chapter regarding impedance matching and standing waves are directly applicable to many problems in optics, such as the transmission through space, and unbounded and bounded media. They also apply to the problems of reflections at interfaces.

8

Refraction and Reflection

The refraction of light was known from the earliest times. Ptolemy discussed refraction and included a table of refractive indices for about 23 substances in his treatise on optics in 150 A.D. Newton employed the nonuniform or dispersive refraction of a prism to demonstrate the constituency of white light. However, it was not until 1850 that Jean B.L. Focault was able to prove that the property of refraction was caused by a difference in the velocity of propagation in the medium.

Refer to Fig. 8-1. A series of parallel waves in the medium on the left marches in and impinges upon a second medium. The medium on the left is less dense; therefore, the velocity of propagation is twice as high as the velocity of the medium on the right. At the interface, some of the incident radiation is reflected and bounces back at an angle equal to the angle of incidence. Without belaboring the point, the angle of incidence and the angle of reflection are identical. Any phase shift that would take place along the plane of the two media would be the same at all points. The reflected wave would therefore set up a standing-wave pattern similar to that shown in Fig. 7-4; however, the amplitude of the reflected component will be less than the amplitude of the incident component since some of the energy will be transmitted.

For the transmitted energy, the slowing causes a bending at the surface, since the wavefronts that are closer in the second medium must match the wavefronts that are farther apart. Figure 8-2 shows this in somewhat more detail. For the incoming wavefront, consider

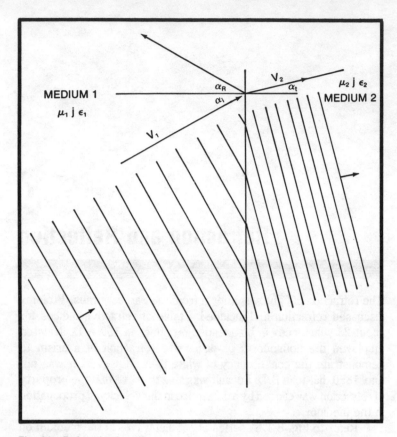

Fig. 8-1. Refractive bending.

points A and B. Point A has just struck the interface in Fig. 8-2. Point B will later strike the interface at point B′. During the time that it takes the wavefront to pass between B and B′ point A will travel a lesser distance A to A′, because the velocity in the denser medium is slower. The two triangles have the side A, B′ in common, so:

$$V_1 = \frac{1}{\sqrt{\mu_1 \epsilon_1}} = A, B' \sin \alpha_1 \qquad \text{Equation 8-1}$$

and

$$V_2 = \frac{1}{\sqrt{\mu_2 \epsilon_2}} = A, B' \sin \alpha_t \qquad \text{Equation 8-2}$$

Equating the two expression gives:

$$\frac{V_1}{\sin \alpha_i} = \frac{V_2}{\sin \alpha_t}$$ Equation 8-3

or

$$\frac{1}{\sqrt{\mu_1 \epsilon_1} \; \sin \alpha_i} = \frac{1}{\sqrt{\mu_1 \epsilon_1} \; \sin \alpha_t}$$ Equation 8-4

The first expression is called *Snell's law of refraction*, propounded by Willebrord Snell in 1621. In the form used by Snell, the velocity was not used. Rather, the relative velocity in the medium with the respect to the velocity of light in free space was used. Presume that V_1 is equal to the velocity of light in free space. Then:

$$\frac{V_1}{V_2} = \frac{C_0}{V_2} = \eta$$ Equation 8-5

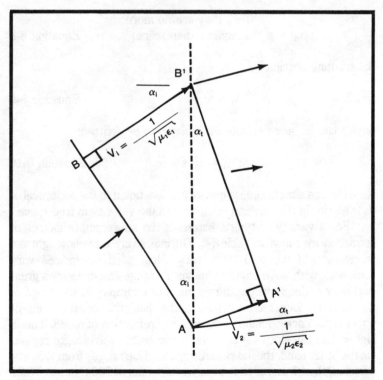

Fig. 8-2. Snell's law illustration.

131

This term is known as the *index of refraction*. This was the quantity actually used in the original statement of Snell's Law and is still the quantity most often used in optical work. The relationship between the velocity of propagation and the permeability and dielectric constant is as shown in Equation 8-1 and 8-2. Thus, substituting Equation 8-5:

$$\frac{V_1}{V_2} = \frac{\sqrt{\mu_2 \, \epsilon_2}}{\sqrt{\mu_D \, \epsilon_D}} = \eta \qquad \text{Equation 8-6}$$

The term e_2 is an absolute term given in farads/meter. The dielectric constant usually quoted in tables is simply a dimensionless ratio compared to the dielectric constant of space:

$$\epsilon_2 = \epsilon_r \, \epsilon_D \qquad \text{Equation 8-7}$$

Now for most optically transparent materials:

$$\mu_0 = \mu_2 \quad \begin{array}{c} \text{(i.e., they are no more} \\ \text{magnetic than space)} \end{array} \qquad \text{Equation 8-8}$$

Substituting obtains:

$$\eta = \sqrt{\epsilon_r} \qquad \text{Equation 8-9}$$

Snell's law for any two materials may also be written:

$$\eta_1 \sin \alpha_i = \eta_2 \sin \alpha_t \qquad \text{Equation 8-10}$$

Note this is a dimensionless quantity proportional to the reciprocal of the velocity in the medium compared to the velocity in free space.

For a variety of typical materials the approximate indices of refraction are shown in Table 8-1. The indices are for yellow light at a wavelength of 0.6×10^{-6} meters. Most of these indices vary somewhat with wavelength in the visible range and can vary a great deal over wider wavelength regions. For example, in the range of wavelengths longer than a meter, the dielectric constant, e_r, of water is 79, corresponding to an index of refraction of 8.888. This is higher than the index for diamond in the visible wavelength region. On the other hand, the index of refraction of ice ranges from 1.86 at a wavelength of 1 meter to 2.04 at wavelengths of 300 meters. While these are both higher than the optical values, they are much closer than the value for liquid water. It is, of course, the slight differences in the index of refraction across the visible range which is responsible for the dispersion of white light into a spectrum. The slight differences in the dispersion of crown and flint glass are used to make lenses achromatic.

ANGLE OF EXTINCTION

Equation 8-10 can be rewritten in the form:

$$\sin \alpha_t = \frac{\eta_1 \sin \alpha_1}{\eta_2} \qquad \text{Equation 8-11}$$

Suppose that the ray originates in the denser medium so that:

$$\eta_1 > \eta_2 \qquad \text{Equation 8-12}$$

Then it can be seen that light cannot be transmitted beyond an angle of incidence:

$$\sin \alpha_1 \leqq \frac{\eta_2}{\eta_1} \qquad \text{Equation 8-13}$$

It may also be seen that violation of this condition would require the sine of the transmitted angle to exceed +1. The limiting angle is called the *angle of extinction*. For water, this angle is 48.59 degrees. This effect may easily be seen when swimming underwater. Outside of the narrow cone, the water surface appears silvery and quite

Table 8-1. Various Indices of Refraction.

GASES OR VAPORS	
Vacuum or empty space:	1
Air	1.0002918
Carbon Dioxide	1.0004498
Mercury Vapor	1.000933
LIQUIDS	
Carbon Disulfide	1.6276
Water	1.3330
SOLIDS	
Ice	1.31
Rock Salt	1.5443
Glass (Crown)	1.48 to 1.61
Glass (Flint)	1.53 to 1.96
Diamond	2.417

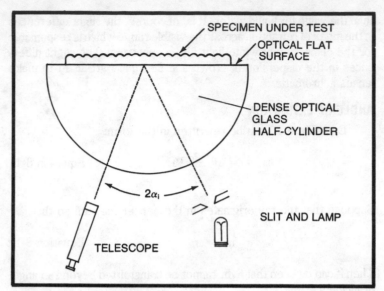

Fig. 8-3. The extinction refractometer.

opaque. For glass with an index of 1.5, the cone is 41.8 degrees. This is the reason that the diagonal of a 45-degree, 45-degree, 90-degree prism will reflect perfectly, even when it is not silvered. When viewed straight in through one of the flat sides, it will appear as if the hypotenuse is mirrored. The line of sight will be turned thru 90 degrees.

One of the first uses for the angle-of-extinction measurement was in determining the index of refraction of different substances. The instrument shown in Fig. 8-3 is often used for determining the index of refraction of liquids, jellies, and other substances. The materials are brought into contact with the optically flat diameter of a cylinder made of very dense optical glass. The angle of extinction is then precisely measured by moving the slit and the telescope until the phenomenon of complete reflection is obtained. This angle of extinction can be found with an accuracy of better than a tenth of a degree using this technique. By the way, the angle of extinction is also very important in the design of optical fibers of some types, as shall be seen shortly.

REFLECTIONS: NORMAL INCIDENCE

As noted earlier (Fig. 8-1), some of the energy incident upon the surface is reflected. Let's next examine the amount of this reflection.

To begin with, assume that the incident radiation is normal to the surface; that is, the angle $\alpha_1 = 0$. In this case, consider that the situation is similar to that of the transmission line of Equation 7-31.

In medium 1, consider that the characteristic impedance is given by:

$$Z_{01} = \sqrt{\frac{L}{C}} = \sqrt{\frac{\mu_1}{\epsilon_1}} \qquad \text{Equation 8-14}$$

This equation is taken directly from the equation of Chapter 7. The substitution for the inductance and capacitance of permeability and dielectric constant is permissible in a homogeneous medium. By a similar token, in medium 2:

$$Z_{02} = \sqrt{\frac{\mu_2}{\epsilon_2}} \qquad \text{Equation 8-15}$$

Then from Equation 7-31 and assuming that the second medium extends an infinite distance to the right, we have:

$$E_b = E_f \left(\frac{Z_{02} - Z_{01}}{Z_{02} + Z_{01}} \right) \qquad \text{Equation 8-16}$$

Using the assumption that all of the μ's are equal, we obtain:

$$E_b = E_f \left(\frac{\dfrac{1}{\sqrt{\epsilon_2}} - \dfrac{1}{\sqrt{\epsilon_1}}}{\dfrac{1}{\sqrt{\epsilon_2}} + \dfrac{1}{\sqrt{\epsilon_1}}} \right) \qquad \text{Equation 8-17}$$

Substituting for the index of refraction and simplifying yields:

$$E_b = E_f \left(\frac{\eta_1 - \eta_2}{\eta_1 + \eta_2} \right) \qquad \text{Equation 8-18}$$

If we assume that medium 1 is free space, then as shown in chapter 7:

$$Z_{01} = \sqrt{\frac{4 \pi \times 10^{-7} \text{ Hy/m}}{1/36 \pi \times 10^{-9} \text{ F/m}}} \qquad \text{Equation 8-19}$$

$$= 120 \pi \text{ Ohms}$$

And by the same token:

$$Z_{02} = \sqrt{\frac{4\pi \times 10^{-7}\text{ hy/m}}{1/36\ \pi \times 10^{-9}\text{ F/m}}} \cdot \sqrt{\frac{1}{E_{r2}}} \qquad \text{Equation 8-20}$$

$$= 120\pi\sqrt{\frac{1}{\epsilon_{r2}}} = 120\pi \cdot \frac{1}{\eta_2} \qquad \text{Equation 8-21}$$

For example, consider the problem of light passing from free space at normal incidence into glass with $\eta = 1.5$ or $e_r = 2.25$.

First calculate the reflected light by the voltage reflection coefficient formula of Equation 8-18:

$$E_b = E_f \left(\frac{1 - 1.5}{1 + 1.5}\right) \qquad \text{Equation 8-22}$$

$$= -0.2 \ \epsilon_f$$

Next note that the voltage on the opposite sides of the interface must be equal. Therefore:

$$E_t = E_f + E_b$$

$$= E_f + (-0.2\ E_f) \qquad \text{Equation 8-23}$$

$$= 0.8\ E_f$$

This requirement for equal voltages on opposite sides of the interface stems from our equivalent circuit. The voltage is actually being measured between two points on the incident side of the interface and two points which are counterparts on the second side of the interface. Each point and its companion on the opposite side are separated by an infinitesimal distance. No voltage difference can therefore be between them.

Nevertheless, the Equation 8-23 seems odd, as though the powers should add up. Therefore, let's see whether the result can be checked against the law of conservation of energy. To begin with, the incident power is:

$$P_f = \frac{E^2_f}{Z_1} = \frac{1}{120\pi}\text{watts} \qquad \text{Equation 8-24}$$

The reflected power is:

$$P_b = \frac{E^2{}_b}{Z_1} = \frac{(-0.2)^2}{120\pi} = \frac{0.04}{120\pi} \text{ watts} \qquad \text{Equation 8-25}$$

And the transmitted power is:

$$P_t = \frac{E^2{}_t}{Z_2} = \frac{(0.8)^2}{120\pi} \times \eta_2$$

Equation 8-26

$$= \frac{0.64}{120\pi} \times 1.5 = \frac{0.96}{120\pi} \text{ watts}$$

The forward power and the reflected power are in medium 1, which has a higher impedance. The transmitted power is in medium 2, which has a lower impedance by a factor of η_2. This all means that the adjustment and power is in fact conserved. The factor of 120π was left in the results to permit an easier comparison. The result that four percent of the incident power is reflected and 96 percent transmitted is in reasonable agreement with empirical results for light on glass in normal incidence.

It is interesting to compare some figures for *diamond*. With $\eta = 2.417$, the reflected power of diamond is 0.172, and the angle of extinction is a mere 24.44 degrees. The very high reflection coefficient and the very small angle of extinction combine for the brilliant reflections obtained from a diamond. The small angle of extinction makes possible the perfect reflections given by shallow cuts. A piece of optical glass cut to the same shape as a diamond will show no "fire" whatsoever.

FINITE THICKNESS SAMPLES: NORMAL INCIDENCE

In the previous example, the second medium was assumed to have had an infinite extent. Therefore, there were no backward waves in the second medium. On the other hand, if the second medium is finite in extent and there is a significant backward wave, the impedance at the first interface can be significantly altered. Also, the relationship between the reflected and transmitted waves compared to the incident wave can be significantly altered. This is the next case to examine.

At any point along a transmission line, or for that matter between the terminals of any circuit, the impedance is given by:

$$Z_x = \frac{E_x}{i_x} \text{ ohms}$$
Equation 8-27

This is simply Ohm's law. It is not necessary that any of the quantities be real, and they might have a phase angle attached to them. It is also not necessary that a pair of terminals be present to which a meter can be attached.

In a bounded medium, such as a two-wire transmission line where a pair of terminals can be attached, the voltage and current can be measured by phase-sensitive voltmeters and ammeters to determine the impedance. However, in an unbounded medium such as plane waves traveling through space, glass, or water, the voltage can be read in terms of volts/meter and the current in amperes/meter. The employment of the transmission line analogy in such cases may distress the purist, but it gives a relatively straightforward picture for visualizing the things that take place. Properly applied, it will also produce the correct answers. For reasons of simplicity, we shall continue its use.

The relationships for the values of E_x and i_x are given by:

$$E_x = E_f \epsilon^{j\omega \left(t - \frac{x}{V} \right)} + E_b \epsilon^{j\omega \left(t + \frac{x}{V} \right)}$$
Equation 8-28

$$i_x = \frac{E_f}{Z_o} \epsilon^{j\omega \left(t - \frac{x}{V} \right)} - \frac{E_b}{Z_o} \epsilon^{j\omega \left(t + \frac{x}{V} \right)}$$
Equation 8-29

These equations can be compared with Equations 7-33 and 7-34. In practice, there is no interest in the ωt component which is in the result. Simplifying to obtain:

$$E_x = E_f \epsilon^{-j\frac{x}{V}\omega} + E_b \epsilon^{j\frac{x}{V}\omega}$$
Equation 8-30

$$i_x = \frac{E_f}{Z_o} \epsilon^{-j\frac{x}{V}\omega} - \frac{E_b}{Z_o} \epsilon^{j\frac{x}{V}\omega}$$
Equation 8-31

In these expressions, the average values of E_x and i_x as a function of position along the transmission line are presented. It should be remembered that the assumptions made earlier in derivation of these expressions precluded any ohmic losses or attenuation. For many of the following examples, these assumptions will prove to yield answers of sufficient accuracy, and the simplification makes the problems much easier to understand.

Referring back to Equation 8-18, we have:

$$E_b = E_f \left(\frac{\eta_1 - \eta_2}{\eta_1 + \eta_2} \right) = E_f \left(\frac{Z_L - Z_0}{Z_L + Z_0} \right) = \Gamma E_f \qquad \text{Equation 8-32}$$

Substituting for E_b yields:

$$E_x = E_f \, \epsilon^{-j \frac{x}{V} \omega} + E_f \left(\frac{\eta_1 - \eta_2}{\eta_1 + \eta_2} \right) \epsilon^{+j \frac{x}{V} \omega} \qquad \text{Equation 8-33}$$

A little bit of algebraic manipulation obtains:

$$E_x = \frac{E_f}{(\eta_1 + \eta_2)} \left[(\eta_1 + \eta_2) \, \epsilon^{-j \frac{x}{V} \omega} + (\eta_1 - \eta_2) \, \epsilon^{+j \frac{x}{V} \omega} \right]$$

Equation 8-34

A similar manipulation applied to the current expression gives:

$$i_x = \frac{E_f}{Z_0(\eta_1 + \eta_2)} \left[(\eta_1 + \eta_2) \, \epsilon^{-j \frac{x}{V} \omega} - (\eta_1 - \eta_2) \, \epsilon^{+j \frac{x}{V} \omega} \right]$$

Equation 8-35

Substituting back into Equation 8-27 gives:

$$Z_x = \frac{E_x}{i_x} = Z_0 \left[\frac{(\eta_1 + \eta_2) \, \epsilon^{-j \frac{x}{V} \omega} + (\eta_1 - \eta_2) \, \epsilon^{+j \frac{x}{V} \omega}}{(\eta_1 + \eta_2) \, \epsilon^{-j \frac{x}{V} \omega} - (\eta_1 - \eta_2) \, \epsilon^{+j \frac{x}{V} \omega}} \right]$$

Equation 8-36

Without belaboring the point, a substitution for the second version of Equation 8-32 would have yielded:

$$Z_x = Z_0 \left[\frac{(Z_L + Z_0) \, \epsilon^{-j \frac{x}{V} \omega} + (Z_L - Z_0) \, \epsilon^{+j \frac{x}{V} \omega}}{(Z_L + Z_0) \, \epsilon^{-j \frac{x}{V} \omega} - (Z_L - Z_0) \, \epsilon^{+j \frac{x}{V} \omega}} \right] \qquad \text{Equation 8-37}$$

Note that the expression $\frac{x}{V}\omega$ is a bit clumsy. If we make the substitutions, $\beta = 2\pi/\lambda$ and $X = -\ell$, the expression can be put in somewhat simpler form. A negative sign has been applied to ℓ because it is measured backward from the load end of the line toward the generator end. With these alterations we can write:

$$\lambda = V/f \qquad \& - \beta\ell = \omega x/V \qquad \text{Equation 8-38}$$

And from Eulers Equation we may also obtain:

$$\eta_1 \left(\epsilon^{\ j\beta\ell} + \epsilon^{\ -j\beta\ell} \right) = 2\,\eta_1 \cos \beta\ell \qquad \text{Equation 8-39}$$

$$\eta_2 \left(\epsilon^{\ j\beta\ell} - \epsilon^{\ -j\beta\ell} \right) = 2\,j\eta_2 \sin \beta\lambda \qquad \text{Equation 3-40}$$

Substituting into Equation 8-36 and canceling the 2's gives:

$$Z_x = Z_0 \left(\frac{\eta_2 \cos \beta\ell + j\eta_1 \sin \beta\ell}{\eta_1 \cos \beta\ell + j\eta_2 \sin \beta\ell} \right) \qquad \text{Equation 8-41}$$

Similarly for Equation 8-37, we obtain:

$$Z_x = Z_0 \left(\frac{Z_L \cos \beta\ell + jZ_0 \sin \beta\ell}{Z_0 \cos \beta\ell + jZ_L \sin \beta\ell} \right) \qquad \text{Equation 8-42}$$

Next, consider some special cases. Suppose that there are three media of impedances: Z_a, Z_b, and Z_c. A plane wave of a single frequency is proceeding from A to C at normal incidence on the AB surface and the BC surface. For the waves, forward and reflected traveling in medium B, the following equation can be written:

$$Z_{ab} = Z_b \left(\frac{Z_c \cos \beta\ell + j\,Z_b \sin \beta\ell}{Z_b \cos \beta\ell + j\,Z_c \sin \beta\ell} \right) \qquad \text{Equation 8-43}$$

Now for the special case where n is any integer and $\beta\ell = n\pi$:

$$\cos n\pi = +1 \qquad \text{for n = 0, 2, 4 \ldots} \infty$$
$$= -1 \qquad \text{for n = 1, 3, 5 \ldots} \infty$$

Equation 8-43 reduces to:

$$Z_{ab} = Z_b \left(\frac{Z_c}{Z_b} \right) = Z_c \qquad \text{Equation 8-44}$$

The conclusion is that the layer of medium B is invisible whenever it is an even number of half wavelengths thick. It has no effect whatever upon the transmission although it does have a standing wave within itself. This effect is wavelength dependent and grows more so as n becomes larger. This principle is often used in interference filters. A transparent medium of high index of refraction is deposited upon a substrate of glass by a process of evaporation in a vacuum. As the deposited layer builds up, the color of light which is transmitted is monitored. When the desired transmission band is attained, the evaporation process can be stopped. The filter has a very high reflection at all frequencies for which it is not an integral number of half waves thick.

Next, consider the case where $\beta\ell = m\,\dfrac{\pi}{2}$, with m any odd integer:

$$\cos\frac{m\pi}{2} = 0 \qquad \sin\frac{m\pi}{2} = \ +1 \text{ for } m = 1, 5, \ldots \infty$$

$$= \ -1 \text{ for } m = 3, 7 \ldots \infty$$

Equation 8-43 reduces to:

$$Z_{ab} = Z_b \left(\frac{j\,Z_b}{j\,Z_c} \right) = \frac{Z_b^2}{Z_c} \qquad \text{Equation 8-45}$$

If medium B is carefully selected so that:

$$Z_b = \sqrt{Z_a\,Z_c} \qquad \text{Equation 8-46}$$

The following is obtained:

$$Z_{ab} = Z_a \qquad \text{Equation 8-47}$$

In this case, medium B acts as a quarter-wave transformer that matches the impedance of medium C to that of medium A; there is no reflection. This is the mechanism used in producing the antireflection coatings seen upon the front lenses of high-quality cameras and binoculars. For glass with an index of refraction of 1.6, a material with an index of 1.26 will provide impedance matching when applied in a quarter-wave layer. Why do these coatings have a purplish color? They are matched in the center of the visible range and reflect rather strongly at the extreme red and blue wavelengths where they are not a quarter-wave thick.

OBLIQUE INCIDENCE

The transmission line analogy breaks down for incidence angles departing appreciably from normal. Referring back to Fig. 8-1, both a backward reflection and a transmitted component exist. The direction of these components is described by Snell's law and the angle-of-incidence equals the angle-of-reflection laws. However, these laws do not describe the proportions between the transmitted and reflected components. The following analysis is directed toward this problem.

In the normal incidence case, it is not necessary to consider polarization because the E (electric) and the H (magnetic) vectors are orthogonal to one another and also to the direction of travel of the plane wave. This means that they will always be parallel to the plane of the interface. For oblique incidence, however, this is not the case so polarization must be considered.

BREWSTER'S ANGLE

Figure 8-4A shows a situation representing a wave obliquely incident upon a second medium. For any polarization of the wave, the wave may be resolved into two components. One of these has the E vector lying in the plane containing the incidence angle, and the magnetic or H vector is parallel to the interface. The other component has the electric vector parallel to the plane of the interface, and the magnetic vector is in the plane of the incidence angle.

For this study of Brewster's angle phenomena, consider the wave component with H parallel to the interface and the E vector in the plane of the angle of incidence. We are concerned here with the wave impedance parallel to the plane of the interface. Therefore, the field components must be reduced to those parallel to the interface. As can be noted in Fig. 8-4B, this amounts to:

$$Z_{1 \text{ tan}} = \frac{E_i \cos \alpha_i}{H_i} \qquad \text{Equation 8-48}$$

$$Z_{2 \text{ tan}} = \frac{E_t \cos \alpha_t}{H_t} \qquad \text{Equation 8-49}$$

The subscript *tan* refers to the tangential component. In any medium:

$$E = \sqrt{\frac{\mu}{\epsilon}} H \qquad \text{Equation 8-50}$$

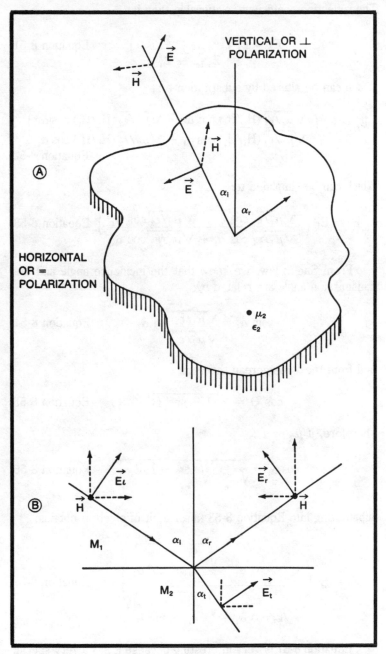

Fig. 8-4. At A, a wave is obliquely incident upon a second medium. At B, the field components must be reduced to those parallel to the interface.

The reflection coefficient relationship then is:

$$E_r = E_i \left(\frac{Z_{2\ tan} - Z_{1\ tan}}{Z_{2\ tan} + Z_{1\ tan}} \right)$$ Equation 8-51

Which can be altered by substitution to:

$$E_r = E_i \left(\frac{\sqrt{\mu_1/\epsilon_1}\,(H_i/H_i)\cos\alpha_i - \sqrt{\mu_2/\epsilon_2}\,(H_t/H_t)\cos\alpha_t}{\sqrt{\mu_1/\epsilon_1}\,(H_i/H_i)\cos\alpha_i + \sqrt{\mu_2/\epsilon_2}\,(H_t/H_t)\cos\alpha_t} \right)$$
Equation 8-52

Which can be simplified to:

$$E_r = E_i \left(\frac{\sqrt{\mu_1/\epsilon_1}\cos\alpha_i - \sqrt{\mu_2/\epsilon_2}\cos\alpha_t}{\sqrt{\mu_1/\epsilon_1}\cos\alpha_i + \sqrt{\mu_2/\epsilon_2}\cos\alpha_t} \right)$$ Equation 8-53

From Snell's law, we know that the incidence angle and the transmission angle are related by:

$$\sin\alpha_t = \frac{\sqrt{\mu_1/\epsilon_1}}{\sqrt{\mu_2/\epsilon_2}}\sin\alpha_i$$ Equation 8-54

And from the Pythagorean theorem:

$$\cos\Theta = \sqrt{1 - \sin^2\Theta}$$ Equation 8-55

Therefore, if $\mu_2 = \mu_1$:

$$\cos\alpha_t = \sqrt{\frac{\epsilon_1}{\epsilon_2}}\sqrt{\frac{\epsilon_2}{\epsilon_1} - \sin^2\alpha_i}$$ Equation 8-56

Substituting into Equation 8-53 (after a bit of algebra) obtains:

$$E_r = E_i \left(\frac{\epsilon_2/\epsilon_1 \cos\alpha_i - \sqrt{\dfrac{\epsilon_2}{\epsilon_1} - \sin^2\alpha_i}}{\epsilon_2/\epsilon_1 \cos\alpha_i + \sqrt{\dfrac{\epsilon_2}{\epsilon_1} - \sin^2\alpha_i}} \right)$$ Equation 8-57

Equation 8-57 is very interesting because it has a very special case built into it. The numerator is the difference of two terms.

When these two terms equal, the numerator must disappear; i.e., no reflection appears at the interface. This case can be solved to be:

$$\alpha_i = \arctan \sqrt{\frac{\epsilon_2}{\epsilon_1}} \qquad \text{Equation 8-58}$$

This is *Brewster's angle*. At this angle, all of the wave component with the H vector parallel to the interface is perfectly transmitted into the second medium because the impedance is perfectly matched.

E VECTOR PARALLEL TO INTERFACE

For the wave component with the E vector parallel to the interface, the H component must be diminished:

$$Z_{1\ tan} = \frac{E_i}{H_{i\ tan}} \qquad Z_{2\ tan} = \frac{E_t}{H_{t\ tan}} \qquad \text{Equation 8-59}$$

By a line of manipulation similar to the previous exercise, the following can be obtained:

$$E_r = E_i \left(\frac{\cos \alpha_i - \sqrt{\epsilon_2/\epsilon_1 - \sin^2 \alpha_i}}{\cos \alpha_i + \sqrt{\epsilon_2/\epsilon_1 - \sin^2 \alpha_i}} \right) \qquad \text{Equation 8-60}$$

This expression is similar to the one for the horizontal H term; however, the absence of the e_2/e_1 term in the first portion of the numerator makes a difference in the behavior. As long as e_2/e_1 is greater than unity, Equation 8-60 simply shows a monotonic decrease in the value of E_r with increasing incidence angle.

Figure 8-5 shows the solution to Equations 8-57 and 8-60. For the Brewster's angle of 56.6 degrees, the light reflected by the second medium would be 100 percent horizontally polarized; furthermore, approximately 16 percent of the horizontal component would be reflected. This is the phenomenon noted by Sir David Brewster. If the curve had been plotted in power, rather than voltage, the minus signs that indicate phasing would disappear.

This Brewster's angle phenomenon is quite commonly visible to someone wearing polarized glasses on a bright sunny day. A number of substances not ordinarily thought of as being optical materials, such as fresh asphalt paving and automobile laquer finishes, will show a very pronounced Brewster's angle effect. At this angle, all of the

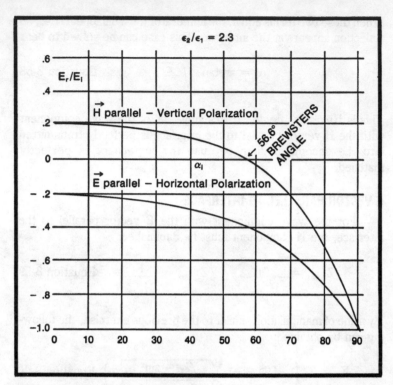

Fig. 8-5. Oblique incidence reflection.

vertically polarized component will be transmitted into the medium and absorbed. Brewster's angle is also very visible on calm water, where it occurs at 53 degrees.

9

Lenses

The first use of glass lenses was to aid eyesight. This usage is very old. Chinese legend tells of the common use of eyeglasses in 500 B.C. Roger Bacon includes information on eyeglasses in his book *Opus majus*, published in 1268 A.D. And Marco Polo reported seeing the Chinese wear eyeglasses during his trip in 1275 A.D. By the time of Galileo, eyeglasses were common throughout Europe. It is to Benjamin Franklin that those of us slightly stricken in years owe the invention of bifocals. In this chapter, we will examine the basic properties of lenses, with particular attention to those types and systems applicable to fiberoptics.

COLLIMATION

One of the most common uses of a lens is to collimate, or render parallel, the light emanating from a point source. The converse operation is to *focus*, or concentrate parallel rays of light to a point, which will be discussed first.

Figure 9-1 represents a group of parallel rays of light incident upon some interface AB. Imagine that the rays are from a star that is so distant that all the rays which strike AB are essentially parallel and that the star is so small in angular extent that it represents only a point. Then concentrate all of the rays from the star into a single point P, which could be on a photographic plate.

All of the waves will be in phase outside the aperture. Inside the aperture, the waves striking the aperture at some height h will be forced to travel some extra distance a to get to point P. To have the

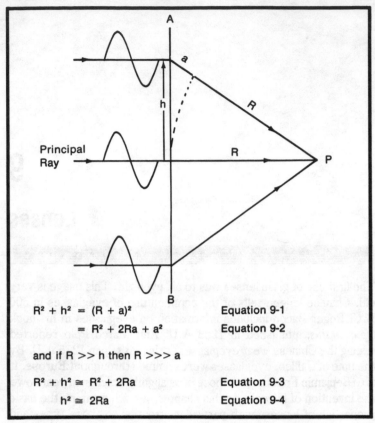

$$R^2 + h^2 = (R + a)^2 \qquad \text{Equation 9-1}$$
$$= R^2 + 2Ra + a^2 \qquad \text{Equation 9-2}$$

and if R >> h then R >>> a

$$R^2 + h^2 \cong R^2 + 2Ra \qquad \text{Equation 9-3}$$
$$h^2 \cong 2Ra \qquad \text{Equation 9-4}$$

Fig. 9-1. Focusing.

energy add at point P, the phase error for the outside waves must be corrected.

Equations 9-1 through 9-4 of Fig. 9-1 show, first of all, the exact calculation of the extra distance from the Pythagorean theorem. However, if the range of h is restricted to a small fraction of R, the term a^2 becomes negligible so that the equation may be simplified, as shown in Equation 9-4. If the ray at h could be speeded up by this amount, the focusing would take place; however, this would require that the ray travel at something faster than the speed of light! This cannot be done practically at optical frequencies. At microwave frequencies, though, a waveguide demonstrates a phase velocity faster than the speed of light in free space. Therefore, this technique is possible and in fact is frequently used for microwave lenses.

In practical optics, however, the usual technqiue is to retard the waves in the central portion of the lens. In Fig. 9-2, the lens has taken the form of a flat slab of thickness t. The slab has a dielectric constant that changes with the height off axis h and is greatest in the center. The design is such that the variation in the index of refraction

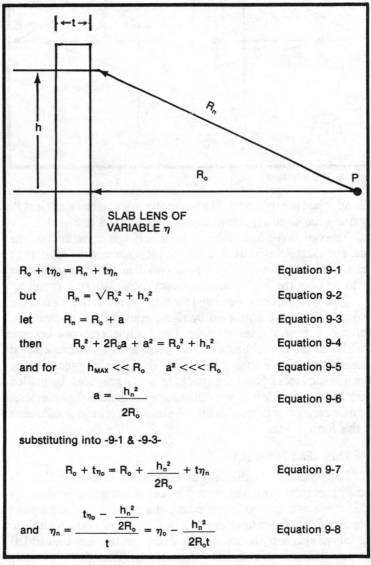

$R_o + t\eta_o = R_n + t\eta_n$ Equation 9-1

but $R_n = \sqrt{R_o^2 + h_n^2}$ Equation 9-2

let $R_n = R_o + a$ Equation 9-3

then $R_o^2 + 2R_oa + a^2 = R_o^2 + h_n^2$ Equation 9-4

and for $h_{MAX} << R_o$ $a^2 <<< R_o$ Equation 9-5

$$a = \frac{h_n^2}{2R_o}$$ Equation 9-6

substituting into -9-1 & -9-3-

$$R_o + t\eta_o = R_o + \frac{h_n^2}{2R_o} + t\eta_n$$ Equation 9-7

and $\eta_n = \dfrac{t\eta_o - \dfrac{h_n^2}{2R_o}}{t} = \eta_o - \dfrac{h_n^2}{2R_ot}$ Equation 9-8

Fig. 9-2. The variable index lens.

149

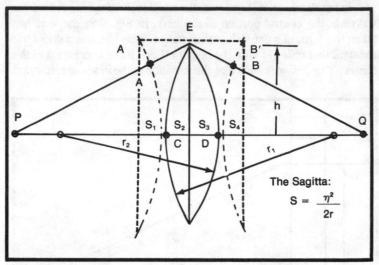

Fig. 9-3. The thin lens.

falls off at just the right rate. Therefore, focusing will take place at P. By using the small angle restriction, Equation 9-8 of Fig. 9-2.

The variation in index is parabolic with height or distance off the axis. For example, $t=0.01\,R_o$ and $h=0.1\,R_o$. Assume the index to be unity. Find the index in the center to be 1.5 and the variation to be parabolic from the axis to the maximum radius where it is unity.

This type lens has been used for many years in the microwave field. It is usually constructed by using graded rings of dielectric material. It is only in recent years that such lenses have become available in the optical field. The developments that have made it possible to fabricate such lenses were the diffusion processes by which graded index fibers are made. In actual practice, the graded index fibers are made first as a rod of some substantial diameter with the index gradation diffused in that. A slice of such a rod would serve as this form of lens.

THE THIN LENS FORMULA

Let's examine the thin lens formula in the format in which it is usually presented in physics texts. Figure 9-3 shows a lens with both sides spherically curved for generality. Light emerges from a point source at P and comes to a focus at point Q. Points P and Q both lie on the axis of symmetry of the system. Points on the arcs CA and DB are equidistant from P and Q, respectively. Therefore, it is only necessary to show the correction for the arc lengths between the spherical wavefronts. For a constant electrical length between P and Q, the following equation exists:

$$AE + EB = \eta(S_2 + S_3) \qquad \text{Equation 9-1}$$

If the lens is very thin and the distance from P and Q are large, little error is involved in using points A' and B' in place of A and B. The projection of these points is equal to the sagitta of the arc. Therefore:

$$S_1 + S_2 + S_3 + S_4 = \eta(S_2 + S_3) \qquad \text{Equation 9-2}$$

Apply the small arc approximation to obtain:

$$\frac{h^2}{2p} + \frac{h^2}{2r_1} + \frac{h^2}{2r_2} + \frac{h^2}{2Q} = \eta\left(\frac{h^2}{2r_1} + \frac{h^2}{2r_2}\right) \qquad \text{Equation 9-3}$$

Which may be reorganized to:

$$\frac{h^2}{2p} + \frac{h^2}{2Q} = \eta - 1\left(\frac{h^2}{2r_1} + \frac{h^2}{2r_2}\right) \qquad \text{Equation 9-4}$$

Which may be simplified to the thin lens formula:

$$\frac{1}{p} + \frac{1}{Q} = (\eta - 1)\left(\frac{1}{r_1} + \frac{1}{r_2}\right) \qquad \text{Equation 9-5}$$

This formula is generally applicable to all thin lenses. For accuracy, it is necessary that the rays to the periphery of the lens subtend small angles from the axis. The drawing has been exaggerated for clarity, and the lens shown would *not* be considered a thin lens.

Note that if the distance from the object to the lens is infinite, the formula becomes:

$$\frac{1}{Q} = (\eta - 1)\left(\frac{1}{r_1} + \frac{1}{r_2}\right)$$

For focal length f, assume P is infinite. Then:

$$\frac{1}{f} = (\eta - 1)\left(\frac{1}{r_1} + \frac{1}{r_2}\right) \qquad \text{Equation 9-6}$$

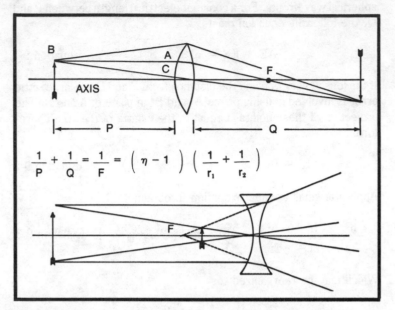

$$\frac{1}{P} + \frac{1}{Q} = \frac{1}{F} = \left(\eta - 1 \right) \left(\frac{1}{r_1} + \frac{1}{r_2} \right)$$

Fig. 9-4. Ray tracings. Convergent lens at A and divergent lens at B.

Stated verbally, the focal length of a lens is the distance at which the lens will image an infinitely distant object.

In Equation 9-6, the focal length is inversely proportional to the radii. If one of the radii becomes infinite, the lens is a *plano-convex* lens. If one of the radii is negative, the lens is a *meniscus* lens. If both radii are negative, the lens is a *dual concave* lens with a negative focal length.

RAY TRACING AND IMAGES

It is often worthwhile to construct a ray tracing of a lens or a lens system. The ray tracing can be done with any number of rays; however, only two are actually necessary.

Figure 9-4A shows a ray tracing for a converging lens. In the ray tracing, the arrow is set up at the object position P. The first ray is drawn from the point of the arrow parallel to the lens axis. It is then drawn from that point through the focus and beyond. The second ray is drawn from the head of the arrow through the center of the lens until it intersects the first ray. The point of the arrow on the image is located by the intersection of the two rays. The ray passing through the center of the lens is not deviated because the sides of the lens are parallel at the center. Here again, the assumption is that the lens is thin. A peripheral ray has been added to the drawing since it is sometimes necessary to know the location of this ray for locating aperture stops and other apparatus.

Figure 9-4B shows the same technique applied to a negative or diverging lens. As noted from the thin lens formula, the focal length for a lens of this type is negative. The image is said to be a *virtual image*, and the focus is said to be a *virtual focus*. This is because it seems to be on the same side of the lens as the object. The diverging lens always makes the image smaller than the object, and a virtual image cannot be projected onto a screen. It can be seen from the far side of the lens from the object, which appears reduced and more distant.

As shown in Fig. 9-4A, the converging lens can project a *real image* upon a screen. This image is inverted with respect to the object. It is possible to obtain a virtual image from a converging lens if the object is inside the focal length of the lens. Figure 9-5 shows this ray tracing. The virtual image is larger than the object and it is erect. This is the mode in which one uses a magnifying glass. The real image is always smaller than the object if the object is more than twice the focal length from the lens. At twice the focal length, the situation is:

$$\frac{1}{p} + \frac{1}{Q} = \frac{1}{f}$$

$$\frac{1}{2f} + \frac{1}{Q} = \frac{1}{f} \quad \therefore \quad Q = 2f$$

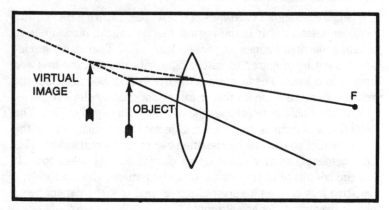

Fig. 9-5. The virtual image from a converging lens.

153

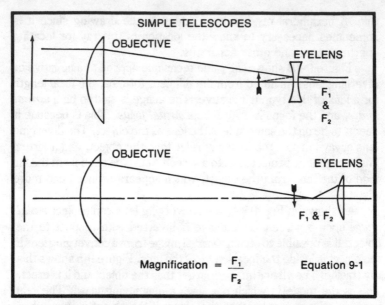

Fig. 9-6. Simple telescopes. The Gallilean telescope is shown at A, and the astronomical refractor is shown at B.

In this case, where the object and image distances are equal, the image sizes are equal. This is the mode in which most photocopiers are operated, producing an image the same size as the object. In the range between twice the focal length and the focal length, the lens can project real images of a size larger than the object. This arrangement is sometimes used in microphotography.

THE SIMPLE TELESCOPE

Figure 9-6 shows two types of telescopes. Figure 9-6A shows a Gallilean telescope. It is the normal function of the human eye to accept collimated images or parallel light rays. Therefore, a telescope must be prepared to gather light with the *objective lens* and bring it into focus. The function of the eye lens is then to render the image collimated again so that it can be focused by the eye.

In the Galillean telescope, the eye lens is a negative lens. The objective lens focus is made to coincide with the virtual focus of the negative lens so that the combination can project a real image. This arrangement is used in cheap opera glasses and toy telescopes. It has the advantage of presenting an upright image. Unfortunately, a negative lens creates a very restricted field of view, especially as higher magnifications are attempted.

154

The second telescope type is sometimes referred to as an astronomical refractor. In this type, the objective lens projects a real inverted image on the focal plane which is viewed through the magnifying eye lens. The image is inverted by projection from the objective. In this type of telescope, the field of view is determined by the design of the *ocular* or the eye lens. The magnification is determined by the ratio of the focal lengths.

CHROMATIC ABERRATION

In actual practice, a telescope as simple as those shown on the illustration would be quite unsatisfactory for magnifications larger than three or four. One of the reasons is the problem of *chromatic aberration*. For crown glass, the index for the C line in the red portion of the spectrum is 1.51263. For the F line of the spectrum in the blue, the index is 1.52080. Applying these to the thin lens formula will yield different focal lengths for the two colors. For example, a simple plano-convex lens which has r_1 equal to 1 meter and r_2 as infinite, would have focal lengths of 0.51263 meters in the red and 0.52080 meters in the blue portion of the spectrum. The red would focus 8.17 millimeters in front of the blue, and it would be impossible to focus an image for red and blue simultaneously.

In fiberoptic telecommunications, this usually poses no problem because the light sources generally have a narrow enough spectrum so that chromatic aberration poses no problem. However, in an endoscope, where imaging is required, this would have to be corrected. The usual technique for providing achromatic correction uses two different lenses that have glass of different dispersive powers. One of the older methods employed crown and flint glass to form two lenses that could be color corrected by correctly choosing focal lengths.

For example, consider the use of crown and flint glasses. These have indices as shown in Table 9-1. Now consider the focal length for

Table 9-1. Crown Glass and Flint Glass Indices of Refraction.

	η C	η F
CROWN	1.51263	1.52080
FLINT	1.61746	1.63466

each of the components at each wavelength:

For Crown

$$\frac{1}{f_{FC}} = (\eta_{FC} - 1)\left(\frac{1}{r_1} + \frac{1}{r_2}\right) \qquad \text{(F line)} \qquad \text{Equation 9-7}$$

$$\frac{1}{f_{CC}} = (\eta_{CC} - 1)\left(\frac{1}{r_1} + \frac{1}{r_2}\right) \qquad \text{(C line)} \qquad \text{Equation 9-8}$$

For Flint

$$\frac{1}{f_{FF}} = (\eta_{FF} - 1)\left(\frac{1}{r_1} + \frac{1}{r_2}\right) \qquad \text{(F line)} \qquad \text{Equation 9-9}$$

$$\frac{1}{f_{CF}} = (\eta_{CF} - 1)\left(\frac{1}{r_1} + \frac{1}{r_2}\right) \qquad \text{(C line)} \qquad \text{Equation 9-10}$$

To match the focal lengths at these two wavelengths, the combined focal lengths of the elements are added:

$$\frac{1}{f_F} = \frac{1}{f_{FC}} + \frac{1}{f_{FF}} \qquad \text{Equation 9-11}$$

$$\frac{1}{f_C} = \frac{1}{f_{CC}} + \frac{1}{f_{CF}} \qquad \text{Equation 9-12}$$

And the focal lengths at the two wavelengths must be made equal:

$$\frac{1}{f_{FC}} + \frac{1}{f_{FF}} = \frac{1}{f_{CC}} + \frac{1}{f_{CF}} \qquad \text{Equation 9-13}$$

Substituting from Equations 9-7 through 9-10 and simplifying gives:

$$\frac{\eta_{FF} - \eta_{CF}}{f_F} = \frac{\eta_{CC} - \eta_{FC}}{f_C} \qquad \text{Equation 9-14}$$

The difference in refractive index is called the dispersive power. For the example given, this is:

$$\frac{0.0172}{f_F} = \frac{-0.00817}{f_C}$$

And the required ratio between the focal lengths of the lenses is:

$$\frac{f_F}{f_C} = -2.105 \qquad \text{Equation 9-15}$$

Thus, if the crown element has a focal length of 1 meter, the flint lens is negative with a focal length of -2.105 meters. And the combination would have a focal length of:

$$\frac{1}{1} + \frac{1}{(-2.105)} = \frac{1}{f} = \frac{1}{1.905 \text{ m.}} \qquad \text{Equation 9-16}$$

This approximate achromatizing formula based upon the thin lens formula is a rather pale version of how a lens designer actually creates an achromatic lens.

SPHERICAL ABERRATION

The spherical shape for the lenses thus far described are used not because they are the best possible shapes but rather because they are the most practical to make with the precision required for optical work. To make a lens, one of the usual techniques is to make use of a tool the same size as the lens blank, or perhaps slightly larger or smaller. This tool is then rubbed over the surface of the lens form with a slurry of abrasive in between. As the tool is rubbed back and forth, it is also rotated. This action is shown in Fig. 9-7. The center of the tool is in contact more often than the periphery. The center is therefore ground away faster. The only geometric shape which will match with this type of variation in orientation is a sphere. For this reason, the surface of the lens naturally turns out to be spherical. Curvatures accurate within less than a millionth of a meter can be generated. The grinding usually proceeds until the correct shape or radius of curvature has been generated with relatively coarse abrasive. As the correct radius is approached, the coarseness of the abrasive is gradually reduced until a very fine finish is obtained. The production of spherical surfaced lenses is thus a process which

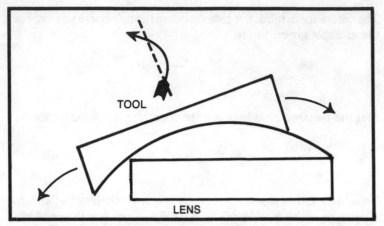

Fig. 9-7. Spherical lens generation. A slurry of abrasive is applied between the lens and the tool as the tool is stroked back and forth while being rotated.

requires a minimum of sophisticated equipment. However, a great many tricks to the art of lens grinding help attain a really satisfactory product. In any event, the use of spherical surfaces for lenses is the result of the relative ease with which such lenses can be made. *Aspheric surfaces* are becoming more common in modern optical devices, but they are very difficult to make and require sophisticated machinery if the curvature departs much from a sphere.

In any event, a spherical surface is often plagued with *spherical aberration*. Figure 9-8 shows a plano-convex lens with r_1 infinite trying to focus collimated light coming from the flat side. This is a worst case representation. (The conventions used in the notation of Fig. 9-3 have been followed.)

Because the side of the lens facing the incoming light is flat and normal to the incoming rays, neglect refraction at this face. Only the refraction at the rear face need be considered. The relationships of Equations 9-1 through 9-4 of Fig. 9-8 are relatively obvious from the geometry of the figure. Because radius r_2 is normal to the glass surface, the derivation of the angle of incidence is relatively obvious. Also, the transmission angle is measured with respect to the radius, so the depression proportional to the difference between the transmission angle and the incidence angle is fairly obvious. The measurements have been referred to the center of the radius of curvature of the rear surface for convenience.

The thin lens formulas imply that the angle of incidence and the angle of transmission are both relatively small. For small angles, the sine is approximately equal to the angle in radians. Substituting in Equation 9-2 of Fig. 9-8 yields:

$$B = \left(\frac{h}{\left(\dfrac{\eta_2}{\eta_1} \cdot \dfrac{h}{r_2} \right) - \dfrac{h}{r_2}} \right) \cos \left(\left(\frac{\eta_2}{\eta_1} \cdot \frac{h}{r_2} \right) - \frac{h}{r_2} \right)$$

Equation 9-17

Simplifying yields:

$$B = \frac{r_2}{(\eta_2 - 1)} \cos \left[\frac{h}{r_2}(\eta_2 - 1) \right]$$ Equation 9-18

Thus Equation 9-5 of Fig. 9-8 can be rewritten:

Focal Error =

$$\underbrace{\left(\frac{r_2}{\eta_2 - 1} \right)}_{F} - \underbrace{r_2 \cos \frac{h}{r_2}}_{A} - \underbrace{\frac{r_2}{(\eta_2 - 1)} \cos \left[\frac{h}{r_2} (\eta_2 - 1) \right] - r_2}_{B}$$

Equation 9-19

In general, $f \cong B$ and $A \cong r_2$. Thus, the thin lens equation is not all that far off. However, the errors are still significant if high performance is expected of the lens. To show this let us, for example, consider a lens in which $r_2 = 0.5$ meters, the lens index of refraction is 1.5, and the maximum value of h is 0.06 meters. For these values, the focal length is 1 meter. Solving Equation 9-19 obtains Table 9-2. For the maximum radius of 60 millimeters, the focusing error exceeds 5 millimeters. The focus for progressively larger radii becomes progressively shorter.

Table 9-2. Focal Errors Versus Wavelengths.

h-meters	Focal Error - meters
0.01	0.00015
0.02	0.00060
0.03	0.00135
0.04	0.00240
0.05	0.00375
0.06	0.00540

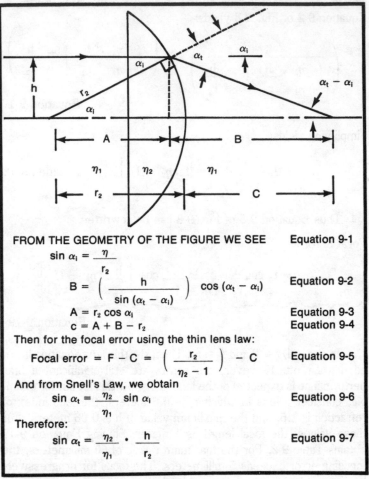

FROM THE GEOMETRY OF THE FIGURE WE SEE   Equation 9-1

$$\sin \alpha_i = \frac{\eta}{r_2}$$

$$B = \left(\frac{h}{\sin (\alpha_t - \alpha_i)} \right) \cos (\alpha_t - \alpha_i)$$ Equation 9-2

$$A = r_2 \cos \alpha_i$$ Equation 9-3

$$c = A + B - r_2$$ Equation 9-4

Then for the focal error using the thin lens law:

$$\text{Focal error} = F - C = \left(\frac{r_2}{\eta_2 - 1} \right) - C$$ Equation 9-5

And from Snell's Law, we obtain

$$\sin \alpha_t = \frac{\eta_2}{\eta_1} \sin \alpha_i$$ Equation 9-6

Therefore:

$$\sin \alpha_t = \frac{\eta_2}{\eta_1} \cdot \frac{h}{r_2}$$ Equation 9-7

Fig. 9-8. Spherical aberration.

Spherical aberration is aggravated when light does not make equal angles with the front and back surfaces of a lens. For minimum spherical aberration, the equal angle with both faces is optimized if the radii of the two faces are arranged as follows:

$$\frac{r_1}{r_2} = \frac{\eta_2 + 4 - 2\eta^2}{\eta_2 + 2\eta^2}$$ Equation 9-20

This relationship between the two radii of curvature minimizes the *circle of confusion*. This statement deserves some explanation. As shown in Fig. 9-9, progressively larger radii tend to provide paraxial rays which cross the axis progressively farther inside the thin lens focus. This gives rise to an envelope or caustic surface which has a minimum diameter at some point. If a star or distant point source of light is focused with a lens having a significant amount of spherical aberration, the image would not focus down to the airy disc and sidelobe pattern, but would have a diffuse disc with a minimum diameter at some best focal length. This best disc focus is the *circle of least confusion.* By a proper choice of the two radii of the lens, the spherical aberration can be minimized, although not eliminated. Some aspheric lens shapes eliminate the spherical aberration; however, these shapes will not necessarily optimize other types of distortion.

ASTIGMATISM

When an object moves off the axis of a spherical lens, another form of aberration is introduced. Figure 9-10 illustrates this effect. The object, which is assumed to be a point source of light, has been moved straight down below the lens axis. In the top view, the point has stayed on axis; therefore, it comes to focus at the normal point Q predicted by the thin lens law. However, the off-axis point in the side view has come to a focus somewhat inside this point at Q'. The figure drawn between these points is as if you took a tube of paper and

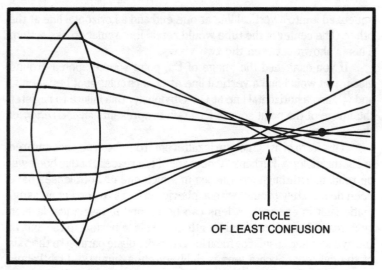

CIRCLE
OF LEAST CONFUSION

Fig. 9-9. The ray paths involved in spherical aberration.

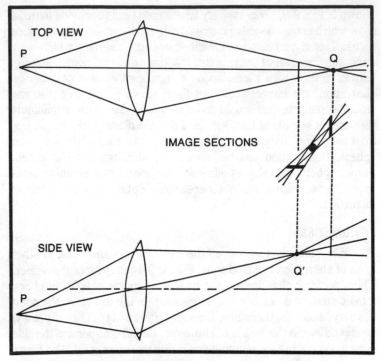

Fig. 9-10. Astigmatism in spherical lenses.

squeezed it into a vertical line at one end and a horizontal line at the other. The center of the tube would retain its circular cross section. This is shown between the two views.

If you examined the image of the point source upon a ground glass, you would find a vertical line at Q, a circular spot between Q and Q', and a horizontal line at Q'. Obviously, this failure to register the image of the point source as a point represents an *aberration* of the system.

This form of astigmatism is different from the type encountered in ophthalmology. Perhaps 20 percent of the readers of this book will be reading it through eyeglasses in which one or both lenses have been deliberately ground with aspheric surfaces to correct an astigmatic fault in eyesight. A lens can be ground in cylindrical fashion such that it has a finite focal length in the plane normal to the axis of the cylinder and an infinite focal length in the plane parallel to the axis of the cylinder. Such a lens is deliberately astigmatic. Ophthalmic lens workers have learned to combine certain amounts of cylindrical

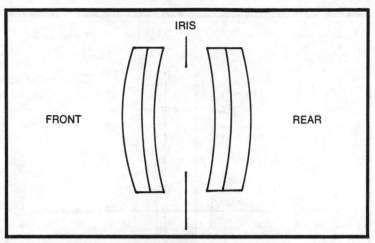

Fig. 9-11. The wide-angle astigmat lens.

"power" with converging or diverging power in the lenses of eye-glasses to correct deficiencies in human eyesight. The astigmatism in these lenses is due to deliberately grinding the lens surfaces to be spherical. In distinction, the type of astigmatism described for the lens of Fig. 9-10 is an aberration of a lens with perfectly spherical surfaces. If the point source had been on-axis, the spot at Q would have been perfectly focused. A lens can be tested for deliberate astigmatism by holding it at arm's length and viewing a distant vertical object. As the lens is rotated about its axis, an astigmatic lens causes the image to tilt away from vertical.

The correction for astigmatism in lens systems is more sophisticated than the correction of either chromatic aberration or spherical aberration. This is particularly a problem in photographic work where wide-angle imaging is required so that the photograph is as sharp on the edges as it is in the center. Figure 9-11 shows one technique used to correct astigmatism of photographic lenses. The lens actually consists of two distinct, separately achromatized elements with concave surfaces turned in toward one another. The iris that regulates light passage is placed between the elements. A mathematical description of the design of such a lens, which simultaneously corrects the major aberrations, is well beyond the scope of this work.

THICK LENSES

The treatment of the various subjects thus far has assumed that the lenses were so thin that there is little error in measuring the

163

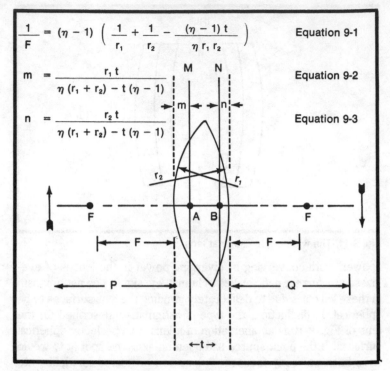

$$\frac{1}{F} = (\eta - 1) \left(\frac{1}{r_1} + \frac{1}{r_2} - \frac{(\eta - 1) t}{\eta \, r_1 \, r_2} \right)$$

Equation 9-1

$$m = \frac{r_1 t}{\eta \, (r_1 + r_2) - t \, (\eta - 1)}$$

Equation 9-2

$$n = \frac{r_2 t}{\eta \, (r_1 + r_2) - t \, (\eta - 1)}$$

Equation 9-3

Fig. 9-12. Thick lens corrections.

object or image distances from either surface of the lens. In practical lenses for fiberoptic work, however, this is seldom the case. The lenses generally used in practical fiberoptic systems usually have a thickness which is a significant fraction of the focal length. For such lenses, the object and image distances must be computed using the *principal points* within the lens. With this one addition, the preceding calculations can be applied. These principal points or *principal planes* can be fairly easily calculated using the following procedure.

Figure 9-12 shows a thick lens. The object and image distances and radii are defined as in Fig. 9-3, with the addition of the principal planes M and N. Equations 9-1 through 9-3 of Fig. 9-12 show the corrections to be applied to the thin lens formulas and the techniques for calculating m and n, the distances to the principal planes from the lens surface on axis. These modifications do not significantly complicate the calculations.

Ray tracing for thick lenses is somewhat modified. Figure 9-13 shows this modification. The parallel ray is drawn to the intersection

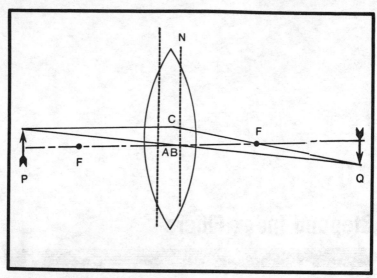

Fig. 9-13. Thick lens ray tracing.

of the rear principal plane N, labeled C in the figure. The ray then proceeds through the focus as before. The central ray is drawn from the tip of the arrow to point A, and then along the axis from point B, such that PA and BQ are parallel. Here again, the additions to the procedure are not too onerous, and a reasonable tracing is quite easily obtainable. After these intersections are found, the peripheral rays are added so that the size of the object is obtained for baffle design and so on.

This is not a lesson in the design of lenses, but rather a familiarization procedure to permit the intelligent use of lenses in fiberoptic systems. In general, the fiberoptic system designer will not design the lenses required by the system. Instead, he will usually purchase these lenses.

10

Stepped Index Fibers

Our discussion of optical fibers will begin with a treatment of stepped index fibers. This makes it somewhat easier to visualize the mechanism of propagation and obtain a "feel" for the properties to be expected of a glass fiber system.

Figure 10-1A shows an example of the cross section of a stepped index fiber. The central material of the fiber is referred to as the *core* and the outer material is described as the *cladding*. The index of refraction of the core is always higher than the index of the cladding. Referring back to our discussion of Snell's law and the angle of extinction formula given in Chapter 8, light cannot be transmitted across an interface where the angle of incidence exceeds:

$$\sin \alpha_1 = \frac{\eta_2}{\eta_1} \qquad \text{Equation 10-1}$$

Suppose that the central core were made up of glass with an index of 1.5 and the cladding were air with an index of 1.0. The angle of extinction would be 41.8 degrees. Figure 10-1B shows that for all angles greater than 41.8 degrees, the light will be internally reflected and will propagate down the glass. For angles less than 41.8 degrees, the light would simply refract outside of the glass into the air. These rays would, of course, simply be lost.

A very important property of the fiber waveguide derives from this relationship. Compare the path taken by the ray which travels

166

down the axis of the fiber with the path taken by the ray at the angle of extinction, which is the last ray that will propagate. It is not too difficult to see that the bouncing ray takes a path equal in length to the hypotenuse of the triangle, whereas the ray which is down the axis of the guide is the base of the triangle. The relative lengths are thus related by the cosine of the angle between the rays. In the case of this example, this angle would be 90 degrees minus 41.8 degrees, or 48.2 degrees, for which the cosine is 0.667.

Over a 1 kilometer path, the axial ray would require:

$$t = \frac{d}{C_o} \eta \qquad \text{Equation 10-2}$$

$$= \frac{1000 \text{ m}}{3 \times 10^8 \text{ m/sec}} \times 1.5 = 5 \times 10^{-6} \text{ sec} \qquad \text{Equation 10-3}$$

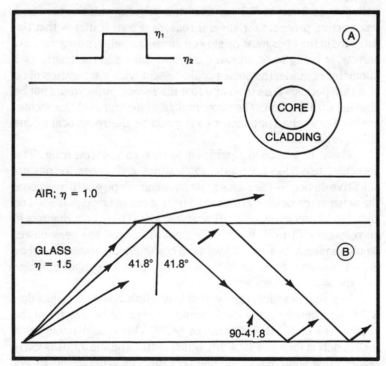

Fig. 10-1. The stepped index fiber. At A, the core and cladding of the fiber. At B, rays with an angle of incidence less than the angle of extinction are lost. Rays with an angle of incidence greater than the angle of extinction will propagate by total internal reflection.

And for the last bouncing ray:

$$t = \frac{\dfrac{d}{\cos \alpha_i}}{C_o} \cdot \eta \qquad \text{Equation 10-4}$$

$$= \frac{1500.3 \text{ m}}{3 \times 10^8 \text{ m/sec}} \cdot 1.5 = 7.5 \times 10^{-6} \text{ sec}$$

You can immediately see that this is not a very good situation. The bouncing ray takes half as much time to arrive as the direct ray. Note that this is due only to the difference in angle between the rays, and not the diameter of the fiber. Obviously, if a 10-ns pulse were launched at one end of the 1-kilometer fiber, there would be a spread of pulses at the far end corresponding to the different modes of propagation. The received energy would be spread out over the period from 5 μs to 7.5 μs after the original pulse. This is an unattractive prospect for several reasons. First, it means that the data rate in the fiber must be slowed down. If a pulse, no matter how narrow, is going to be smeared out over 2.5 μs in the transit of 1 kilometer of guide, the pulses could be sent with a separation of no less than perhaps 3 μs to assure that the second pulse would not be jumbled with the first. The maximum data rate that could be accommodated over the 1-kilometer path would be the reciprocal of this delay, or 333 kHz.

There is a second significant feature to the smearing. The detectors used for fiberoptic applications are generally power-sensitive devices—the signal must have enough power to overcome the noise in the device. Let us assume that our original pulse at 1 ns contained 1 picojoule, or 10^{-12} watt-seconds. Over 1 ns, that would correspond to 1 mW. In spreading out to 2.5 μs, the peak power would diminish to 4×10^{-7} watts. Thus the peak power would be diminished by a factor of 2500, or 34 dB. This is a significant bite into the system sensitivity.

In actual practice, the system bandwidth could be reduced to 333 kHz. Therefore, the noise would be reduced by a factor of the square root of 10^9 Hz/333 kHz, or 54.77. This would amount to a loss in detectability of 17.4 dB, which is still significant. It is fairly obvious that some technique that prevents smearing to this extent would be worthwhile in the overall system considerations.

One thing which is shown by the examination thus far is that the amount of pulse smearing is a time-based consideration, whereas the data rate is a reciprocal time function. Since the amount of time smearing is proportional to the length of the line, the permissible data rates are proportional to the reciprocal of the line length.

PROPAGATION MODES

There is another item which makes the air-clad line unattractive. If you consider that the extinction refractometer shown schematically in Fig. 8-3 will operate at all, it becomes apparent that the electric fields do not abruptly cease at the interface between the high-index material and the low-index material. This fact can also be inferred from the discussion on wave phenomena and interference. For some significant fraction of a wavelength, the electric field must extend outside the core. This nonpropagating extension is termed an *evanescent mode*. This evanescent mode would couple into a propagating mode in any speck of dust that adhered to the surface of the guide, the supports, etc. Therefore, it is worthwhile to supply some cladding that provides a physical support and a containment of all of the evanescent modes so they are isolated from dirt and support contact structures.

The *delay dispersion* in this type of guide could be minimized by making the extinction angle as large as possible. From Equation 10-1, you can see that this could be accomplished by making the two indices as close as possible. In a practical fiber, the core has an index of 1.48 and the cladding has an index of 1.46. This would yield an extinction angle of 80.57 degrees. The cosine of 90 degrees minus 80.57, or 9.43 degrees, is .9865. The reciprocal of the cosine is 1.0137. For the 1-km example, then, the direct ray would require:

$$t = \frac{1000 \text{ m}}{3 \times 10^8 \text{ m/sec}} \times 1.48 = 4.93 \times 10^{-6} \text{ sec} \quad \text{Equation 10-5}$$

And the bouncing limit ray would require:

$$t = \frac{\dfrac{1000}{.9865}}{3 \times 10^8 \text{ m/sec}} \times 1.48 = 5.0 \times 10^{-6} \text{ sec} \quad \text{Equation 10-6}$$

For a net difference of:

$$t = 67.58 \times 10^{-9} \text{ sec}$$

The reciprocal of this is 14.8 MHz. This is a great improvement over the previous situation, both in terms of smearing of the pulse and in shortening of the pulse power peak. The difference of about 1.4 percent in the indices seems to represent a practical limit in the fabrication of the glass and also in the *acceptance cone angle,* which will be discussed shortly.

Thus far, the simple ray discussion has had no mention of the diameter of the fiber, whereas the discussion of waveguides in Chapter 7 noted that there were actually some very critical relationships between the size of the guide and the size of the waves being guided. In particular, the seawall analogy of Fig. 7-4 showed that no waveguide of any type can support a propagating transversely polarized mode if the dimensions are smaller than a half-wavelength. Figure 7-4 also showed that the number of modes which can be supported is proportional to the cross section measurements of the guide in half-wavelengths. In practice, the energy in a stepped index fiber is not propagated at all incidence angles between the extinction angle and 90 degrees. Instead, the energy is propagated in a number of waveguide modes which would correspond to a number of discrete incidence angles in the ray-optic analogy.

The actual development of the mode theory from Maxwell's equations is a formidable task and will not be attempted here. However, several of the results are interesting. One of these is the *mode volume* of a given guide. The number of modes which can propagate in a given guide is given by:

$$N = \frac{V^2}{2}$$ Equation 10-7

The term V is called the *normalized frequency* and is given by:

$$V = \frac{\pi d}{\lambda} \sqrt{\eta_1^2 - \eta_2^2}$$ Equation 10-8

where λ = wavelength, and d = diameter of the core in same units.

If the fibers of the previous example had a core diameter of 50 $\times$ 10^{-6} meters and the wavelength were 0.89×10^{-6} meters:

$$V = \frac{\pi \times 50}{0.89} \sqrt{(1.48)^2 - (1.46)^2} = 42.8$$ Equation 10-9

$$N = \frac{(42.8)^2}{2} = 915 \text{ modes}$$

170

Most of these modes will have a different group velocity. Therefore, a further reduction in the dispersion can be accomplished only by a reduction in the number of modes.

A case of particular interest exists when V is smaller than 2.405. In this case, only a single mode can exist within the guide if polarization moding is neglected. For the indices previously noted, this would require a core diameter of 3.55×10^{-6} meters. Such a fiber is referred to as a *single mode* fiber. It will propagate only one mode and therefore shows no *modal dispersion* at all. Unfortunately, this extremely fine core poses great alignment problems. The core diameter is only one-eleventh of the diameter of a fine blond hair.

A first suspicion might be that it would be worthwhile to go to longer wavelengths. Figure 10-2 shows the attenuation-versus-frequency plot for a representative stepped index fiber. In this case, the fiber has a core diameter of 100 microns. The minimum attenuation is in the vicinity of 0.830 microns. A pronounced absorption peak is around 950 microns, which is attributed to hydroxyl radicals in the glass. Thus, the 0.89-micrometer wavelength is about the largest that can be used efficiently with this make of fiber. In general, most fibers will show a series of absorption peaks beyond approximately 0.9 micrometers.

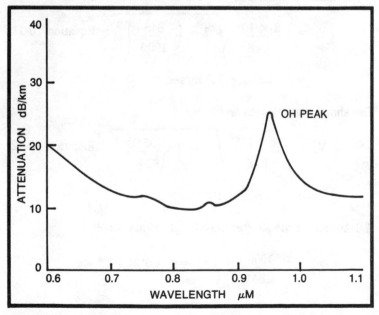

Fig. 10-2. Attenuation of stepped index fibers.

Most glass fibers show a minimum of attenuation between 0.8 and 0.85 micrometers and a second somewhat lower window near 1.05 micrometers. Unfortunately, the LED sources commonly used generally peak in the 0.8 to 0.850 micrometer range.

The few available sources in the 1.05-micrometer range are rather low in power by comparison, thereby restricting their use. The most common LED and ILD sources operate with a peak output in the 820-nanometer range (0.82 micrometers). This wavelength is well below the 730-nanometer point which marks about the bottom of the visible wavelength spectrum in deep red.

Although a single-mode fiber has no intermodal dispersion, it will still suffer somewhat from intramodal dispersion and material dispersion. You have already seen that the velocity of propagation in a single-mode also has a variation with frequency or wavelength given by:

$$V_9 = V_0 \sqrt{1 - (\lambda_0/\lambda_c)^2} \qquad \text{Equation 10-10}$$

For example, a waveguide has a cutoff wavelength of 1900 nm and a LED source has its output spread between 795 nm and 845 nm. Assume that the index of refraction is 1.48, measured in the center of the emission band. The longest wavelength would be:

$$V_9 = \frac{3 \times 10^8}{1.48} \sqrt{1 - \left(\frac{845}{1900}\right)^2} \qquad \text{Equation 10-11}$$

$$= 1.816 \times 10^8 \text{ m/sec}$$

The shorter wavelength would be:

$$V_9 = \frac{3 \times 10^8}{1.48} \sqrt{1 - \left(\frac{795}{1900}\right)^2} \qquad \text{Equation 10-12}$$

$$= 1.841 \times 10^{-8} \text{ m/sec}$$

A 1-km run at the shorter wavelength would have:

$$\frac{1000}{1.841} \times 10^{-8} = 5.432 \times 10^{-6} \text{ sec}$$

And at the longer wavelength:

$$\frac{1000}{1.816} \times 10^{-8} = 5.508 \times 10^{-6} \text{ sec}$$

This would amount to a difference of 76 nanoseconds per kilometer.

Note that this intramodal dispersion is a function of the proximity to the cutoff wavelength. As the transmitted wavelength gets smaller and smaller with respect to the cutoff wavelength, the value of the radical gets closer and closer to unity and the intramodal dispersion gets smaller. This example would be for a strictly single-mode fiber. By comparison, a fiber with a 50-μm core would have a cutoff wavelength on the order of 50 μm and the intramodal dispersion would be only about 4×10^{-11} sec., although the intermodal dispersion on this fiber would be on the order of 69 nanoseconds.

In general, the *material dispersion* works in the opposite direction, with most materials showing a slightly lower index of refraction for the longer wavelengths. If the core material for the fiber were to show the same dispersive power at 820 nanometers as crown glass in the visible region, the indices would be 1.477 at 845 nm and 1.480 at 745 nm. This difference would yield a dispersion of 10 nanoseconds with the long-wave components arriving first.

MODE COUPLING

If the fibers were absolutely perfect, no coupling would be between modes in the fiber. Only those modes which were launched would propagate. At microwave frequencies in hollow metal pipe waveguides, it is possible to maintain very high orders of mode purity in long, straight waveguide runs. A guide which could support perhaps 5 or 7 modes may be excited in only one of the modes. That mode will still account for 99 percent of the power at a kilometer away. However, any bending of the guide—a single bad joint—will convert some of the energy into some of the other modes.

It is usually not possible to identify the modes present in optical fiber guides. The frequencies are too high to permit a direct measurement of the electric fields or magnetic fields. Also, the individual modes are too small to permit direct viewing with optical instruments because of the airy resolution limit. This limit states that no instrument can resolve anything less than a few of the operating wavelengths.

Although you cannot resolve the individual modes, you do know that coupling exists between the modes, because as a fiber grows longer, the intermodal dispersion ceases to grow linearly with fiber length. It has been shown experimentally that for short fibers:

$$\text{Dispersion} = KL \quad \text{nsec/km} \qquad \text{Equation 10-13}$$

And for lengths greater than the coupling length L_c:

$$\text{Dispersion} = K \sqrt{LL_c} \quad \text{nsec/km} \qquad \text{Equation 10-14}$$

Mode coupling effects can be attributed to bending of the fiber, stresses in the fiber, minute imperfections in the fiber, and other inhomogeneities. Unfortunately, this mode coupling also increases the losses because it is usually accompanied by conversion of some of the energy into modes which propagate in the cladding. These modes are generally lossy since the material used to cover the cladding is generally a plastic of some sort which does not have the same properties as the glass.

The coupling length is generally not specified in the data sheets supplied by fiber manufacturers. It is generally greater than 1 kilometer. It may be that the coupling length, being largely attributable to imperfections in the fiber and bending (to some extent), may not be sufficiently constant from fiber to fiber to permit specification on a data sheet. Also, there is not sufficient data to permit an evaluation of the effects caused by fiber curvature. Multiple-kilometer lengths of fiber are usually measured in the laboratory with the fiber on a spool. In the field, the fiber is generally relatively straight; however, it is often subject to some stresses other than the spool coiling. The temperature is seldom the same and so on. It is therefore difficult to obtain good correlation between the coupling length determined in the laboratory and installed guides.

The measurement of dispersion is shown in Fig. 10-3. An electrical circuit drives the source that injects light into the guide. After some distance of travel over the guide, the output pulse is detected from the guide. The output of the detector is processed in electrical circuitry. All of the elements in this arrangement generally tend to degrade the pulse width. The original pulse formed by the drive circuitry is broadened somewhat by the finite rise time of the optical source. Time dispersion is generated by the length of the fiber. At the receiving end, the detector has a finite rise time. The electrical circuitry at the receiving end also has a finite rise time. In

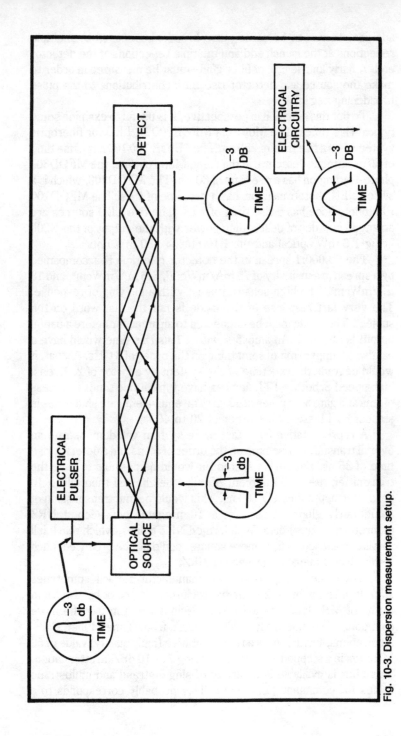

Fig. 10-3. Dispersion measurement setup.

addition, there can be some pulse broadening caused by multiple reflections at the launch end and multiple reflections at the detector end. A fairly long length of fiber guide must be measured in order to make the source and detector rise time contributions to the pulse broadening negligible.

To put this situation in perspective, it is useful to examine some typical rise times. The Motorola MFOE200 IR LED for fiberoptic systems has a rise time of 250 ns. The MFOE 100 100 has a rise time of 50 ns. Among detectors in the same Motorola line, the MFOD300 photo-Darlington has a rise time of 60 μs. The MFOD200, which is a silicon NPN phototransistor, has a rise time of 4 μs. The MFOD100 PIN photodiode has a rise time of 1 ns. The two LED sources are arranged in order of descending power with the output of the E200 being 1.6 mW optical and the E100 being 550 mW optical.

The 60,000:1 spread in the detector rise time is accompanied by a spread in sensitivity of 75 mA/mWcm^2, 5.6 mA/mWcm^2, and 18 μA/mWcm^2. The high sensitivites go with a slowing of response. The very fast response of the diode is not the last word on the subject. This diode must be connected to an amplifier before a useful output is obtained. An amplifier with a 1-ns rise time would have a bandwidth upper limit of something on the order of 3 GHz. At that, it would degrade the rise time of the system by a factor of 2. Even a high-speed Schottky TTL devices have typical rise times in excess of 1 ns and gate shift times of 12 ns. The equivalent gate shift times in standard TTL are on the order of 20 to 40 ns.

A representative fairly fast pulse system will demonstrate an overall transmitter rise time on the order of 50 ns and a detector rise time of 35 ns. For such a system, it would be possible to have the transmitter rise to 90 percent of pulse height and then fall to 10 percent in approximately 100 ns. This would permit transmission of some fairly sloppy pulses at a 10 megabit per second NRZ (nonreturn-to-zero) data. Well formed NRZ pulses, which reach full amplitude and look a little more square, could not be produced much faster than 3 megabit per second NRZ.

When specifying dispersion, manufacturers will sometimes state it in terms of ns/km, or the reciprocal factor of bandwidth in terms of MHz-km. The MHz-km specification is in fact the more common. A commercially available fiber from Times Fiber Communications, Inc. features a representative (high-quality) value of 50 MHz-km in a stepped index fiber offering 7 or 10 dB/km attenuation. This fiber is available in a variety of single-strand and multistrand cables for communications work. This probably corresponds to a dispersion of 7 to 10 ns/km. The lack of standardization when specifying this parameter is a source of minor difficulty in system design. The lack of standardization can lead to a 2:1 or more error in the estimation of the usable bit rate if the specification is given in MHz-km. Or it can lead to a similar estimation error of the bandwidth of a video system if the specification is given in ns/km.

Add to this the coupling length difficulty. Ironically, this coupling length difficulty is prone to become greater as the overall quality of the fiber processing rises along the learning curve. It seems likely that fiberoptic processing will follow the pattern set in silicon device processing with a gradual decrease in imperfections in the product due to experience in volume production and improvements in processes and processing machinery. This should be followed by a decrease in attenuation and an increase in coupling length for the improved, more uniform fibers.

NUMERICAL APERTURE

A very significant feature of the fiber may be deduced from the total internal reflection model shown in Fig. 10-1. Figure 10-4 shows an enlarged view of one of the ends of the fibers. Presume that this end is optically flat due to either a cleaving or polishing operation. The last totally reflected ray is given by subtracting the angle of extinction from 90 degrees. When the ray strikes the interface, it is refracted according to Snell's law to a wider angle caused by the lower index of the air outside the fiber. The definitions for the numerical aperture (NA) and the acceptance cone half-angle are shown in Fig. 10-4.

This is a very significant feature of the fiber because it shows that a ray striking the fiber end at an angle wider than the acceptance cone half-angle will either be bound in a lossy cladding mode or else be lost on the first bounce. In the previous example, where a core index of 1.48 and a cladding index of 1.46 were used, the extinction angle is 80.57 degrees, and the greatest axial angle for a propagating ray is 9.43 degrees. This yields a NA of 0.2425 and a half-cone angle of 14.03 degrees. This value is typical of low-loss stepped index fibers.

This acceptance half-cone angle is not related to the diffraction limited angle discussed earlier in Chapter 6. For example, the first zero angle for a uniformly illuminated circular aperture was given to be 1.14 λ/a radians. For a 50-micrometer fiber and a wavelength of 0.82 micrometers, this would give a first zero at an angle of 0.0187 radians, or 1.07°. The diffraction effect is not negligible; however, it

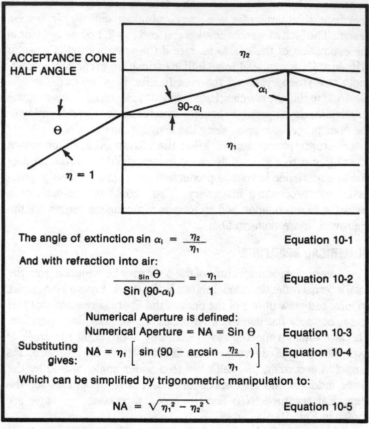

ACCEPTANCE CONE HALF ANGLE

η_2

α_i

$90 - \alpha_i$

Θ

η_1

$\eta = 1$

The angle of extinction $\sin \alpha_i = \dfrac{\eta_2}{\eta_1}$ Equation 10-1

And with refraction into air:

$$\frac{\sin \Theta}{\sin (90 - \alpha_i)} = \frac{\eta_1}{1}$$ Equation 10-2

Numerical Aperture is defined:

Numerical Aperture = NA = $\sin \Theta$ Equation 10-3

Substituting gives: $NA = \eta_1 \left[\sin \left(90 - \arcsin \frac{\eta_2}{\eta_1} \right) \right]$ Equation 10-4

Which can be simplified by trigonometric manipulation to:

$$NA = \sqrt{\eta_1{}^2 - \eta_2{}^2}$$ Equation 10-5

Fig. 10-4. Numerical aperture.

is small with respect to the mode angles for this multimode fiber.

For the single-mode fibers, though, the case becomes somewhat different. If the core diameter is shrunk to 2.59 microns for a single-mode fiber, the full angle for the diffraction pattern becomes 0.361 radians, or 20.68 degrees. NA is 0.353. The diffraction effects come to dominate the control of the acceptance angle for very small fibers.

The acceptance angle is particularly important because most of the solid-state sources do not produce the same well collimated beam produced by gas lasers. A significant spread in the angle of the radiation is characteristic of the radiation produced by these devices. The portions of the radiation which lie outside of the acceptance cone are lost and cannot be used. A large numerical aperture is thus able to capture more radiation from sources with a wide dispersion angle.

The effects of mode coupling enter into the acceptance angle and the NA of optical fibers. An acceptance angle that is measured on a short length of fiber will be larger than the same parameter measured with a length of fiber approaching the coupling length. Figure 10-5 shows an arrangement for measuring the acceptance cone angle of a fiber.

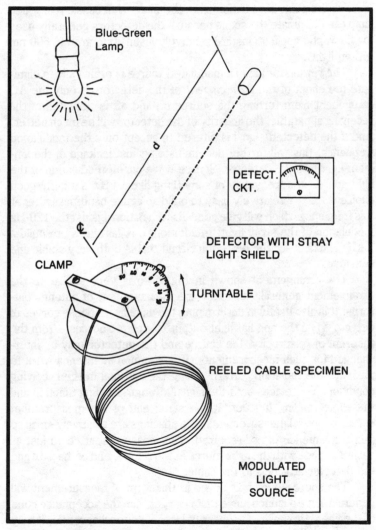

Fig. 10-5. Acceptance cone angle measurement.

In Fig. 10-5, the fiber to be measured is clamped above a calibrated turntable so that the end is approximately above the center of rotation. A detector is placed at some distance d such that the subtended angle of the detector light gathering objective is negligible with respect to the cone angles to be determined. A modulated light source is applied to the end of the reel of fiber cable. The distant detector records the received light as a function of the rotation angle of the turntable. The record is usually normalized to unity at the peak of radiation from the fiber. If necessary, a green lamp can illuminate the scene because the detectors generally used for fiberoptic applications are relatively insensitive around 500 nm (green light).

The reason for using a modulated source is principally to eliminate the effect of the *dark current* of the detector. If only the AC component from turning the source on and off is detected in the receiving apparatus, the linearity of the detector will be much better. And if the detected signal is filtered to accept only the modulation frequency, this will further discriminate against leakage of the ambient light into the detector. Square wave on/off modulation of the light source at a frequency of something like 1 kHz is a fairly good choice since it is relatively easy to build an active band-pass filter at this frequency which will give good discrimination against the 120-Hz modulation of the stray light. In addition, it is fairly easy to build a 1-kHz amplifier for the detection circuit that is both very stable and sensitive.

The arrangement shown in Fig. 10-5 is very similar to the arrangement generally used for the measurement of antenna patterns. It is also useful in determining the angular response curves of detectors and the angular light output curves of sources. From the theorem of reciprocity, the source and the detector may be interchanged for such measurements. It is common in antenna work to measure the radiation patterns using the antenna in the receiving condition. This places both the signal strength reading function and the angle reading function at the same end of the measurement setup. In optical measurements, the spacings are relatively small so this convenience is not required. The measurement could just as easily be made with the light source at the remote end of the path and the detector attached to the cable.

The sources which are used in this form of measurement will generally have a launch angle that is larger than the acceptance cone of the fiber. However, for fibers of large numerical aperture, it may be necessary to employ a source which has a large launch angle in order to assure that the cable alone restricts the acceptance cone. If a narrow launch angle source is eventually to be used in the system, a second measurement of the system that employs the intended source might be required. This is because the output acceptance cone can be modified by a source angle which is smaller than the acceptance cone of the cable itself.

Figure 10-6 shows the acceptance cone measured on a commercially available plastic fiber link for which transmitter, receiver, and a length of plastic cable are supplied. The acceptance cone is not a sharp cornered "slice of pie" or circle segment. The corners are rounded and the main lobe follows a multiple cosine function which has been fitted on the curve of data points. Also, at the wide angles in the figure, the light does not really fall to zero. There are probably two reasons for this. First, you would expect from diffraction theory that there would be wide-angle sidelobes, as shown in Fig. 6-10. However, these should be of lower level than shown on that plot. Secondly, the optical finish on the end of the fiber is something less than perfect and a wide-angle scatter can be expected from each

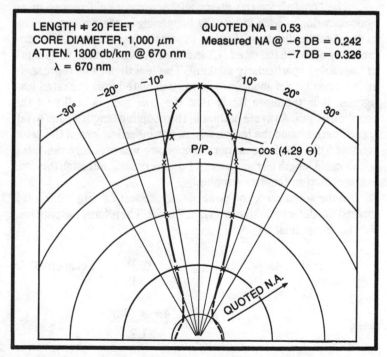

Fig. 10-6. Measured acceptance cone on Dupont Crofon® cable.

scratch and imperfection in this surface. This is the source of the major portion of the wide-angle scatter seen in Fig. 10-6. It is assymmetric and only approximately rendered in the figure.

The rounding of the shoulders of the main beam is itself probably caused by a combination of several factors. The plastic cable itself is a moderately lossy one, and the wide-angle modes are likely to be considerably more lossy than the central modes. The wide-angle radiation emerging from the end of the fiberguide is therefore further attenuated. Secondly, the power launched from the source is not a sharp-cornered affair either. It was not possible to produce a similar plot from the source itself. The mechanical construction of the fitting tended to vignette the pattern at relatively shallow angles. In the plastic fiber, the surface interface between core and cladding is probably not as optically smooth as it is in a glass fiber. Therefore, the attenuation per bounce is probably greater and the wide-angle modes would probably suffer greater attenuation than the shallow-angle modes, which would experience fewer bounces. This effect most likely contributes to higher attenuation per unit length of the plastic fiber.

The "cones" for the fibers and sources really represent an approximation of what is actually happening in the apparatus. This matter will be discussed further in subsequent chapters. The discrepancy between the rated NA and the NA as measured on a length of the cable is particularly striking. The length of cable represents an insertion loss of about 7.93 dB, as calculated from the rated loss figures. It is therefore likely that the specimen is well past the coupling length. As a rule of thumb, the coupling length seems to fall somewhere around the length where the insertion loss of the cable is about 5 dB. The large numerical aperture was probably measured with a cable length on the order of 1 meter or less, although this was not specified in the data supplied.

In the setup used, the *throw* designated as d in Fig. 10-5 is 6.2 cm and the detector active area is 1.7 mm². Therefore, the path loss for the setup in air is:

$$\text{Path Loss} = 10 \log_{10} \frac{4\pi\ d^2}{ad} \qquad \text{Equation 10-15}$$

$$= 10 \log_{10} \frac{4\pi \times (62)^2}{1.7}$$

$$= 46.84 \text{ dB}$$

The numerator in the upper fraction is simply the area of a sphere with a radius equal to the throw in the measurement. The receiver could be made to provide a strong detected signal at this insertion loss. This fact indicates that the system would have functioned satisfactorily with a cable of 46.48 + 7.93/1300 dB/km, or 41.8 meters (159 feet) in length. The operation of this system with a 20-foot cable, even using the relatively lossy plastic, is obviously conservative. With the path loss of the pattern measurement setup, the signal to noise ratio of the system is in excess of 25 dB.

The acceptance cone angle is something that should be measured in a given system if an accurate value of the working numerical aperture is to be obtained. For systems using the much lower loss glass fibers, the coupling length is of course much greater and a working system can be shorter than the coupling length.

11

Graded Index Fibers

Of all of the materials used by man, glass is one of the most widely used and versatile. In tension, glass is stronger than steel and is extremely rigid. Glass has a low coefficient of thermal expansion and is a very good electrical insulator. It is resistant to nearly all acids and bases, except the fluorides, and can be used as a nonreactive container with nearly infinite lifetime.

Long before man used bronze or iron, he had learned to use glass. Glass is made by natural processes in two principal ways. When lightning strikes sand, the heat of the electrical discharge sometimes fuses the sand into elongated branched objects called *fulgurites*. This glass is generally of relatively nonuniform properties and is not particularly useful. The heat and pressure of a volcanic eruption can fuse rock and sand into a hard mineral called *obsidian*. The trace elements in obsidian make it nearly always black, although it may have red streaks. While obsidian is not transparent, it is extremely hard and strong. It was very useful to primitive man.

Very early man found that razor-sharp knives, arrowheads, and spear points could be chipped from obsidian by using the simplest of tools, such as a sharp rock. Obsidian objects have been found in archeological digs around the world. Some of the knives chipped from obsidian rival the sharpness of today's finest scalpels. The corrosion resistance of the glass is so great that these tools survive to this day. The American Indian was also a user of obsidian knives, daggers, and spear points. The black, shiny surface of obsidian also made it a logical choice for jewelry and adornment. Examples of paleolithic obsidian jewelry are also relatively common.

The origin of man-made glass is lost in antiquity. However, it seems likely that this discovery occurred in either Egypt or Syria sometime before 3000 B.C. By 1500 B.C., glass was being pressed in Egypt into bottles, goblets, and other containers. Glassware from the tombs of the Pharaohs may be seen on display in the glass museum in Corning, NY. From about 1500 B.C. to 300 B.C., Egypt was the center of glass manufacture in the world. The bottles, vials, goblets, plates, and other objects were made with a remarkable craftsmanship.

The blowpipe was invented sometime around 300 B.C. This brought forth a further flowering of the glassworking arts. At about the time of the birth of Christ, the first transparent sheet glass was produced in Rome. The quality of this glass must have left something to be desired, however, because Saint Paul wrote to the Church at Corinth of "seeing as through a glass; darkly" (*1 Cor.* 13:12). One of the most celebrated pieces of glass to survive from this era is the Portland Vase, now housed at the British Museum. This richly decorated vase was made in Rome at about 70 A.D. and is reputed to be the most valuable piece of glassware in the world.

Glassmaking came to America in 1608 in Jamestown, VA. This effort failed within the first year, but the Jamestown colonists tried glassmaking again in 1620. This effort was wiped out by the Indian massacre of 1622. The actual start of a continuous glassmaking industry in America dates to 1739 when Caspar Wistar built a glass plant in Salem County, NJ. This operation lasted until 1781 when a great many competitive operations had sprung up. The early American windows were made by blowing a large glass bubble which was cut open and allowed to sag flat on a surface. This process left a crown or thickened lump in the center of the pane. Even today, visits to colonial New England houses will reveal many of these handblown window panes of crown glass.

MAKING GLASS

The basic materials from which glass is made are sand, soda ash (sodium oxide), and lime (calcium oxide). Soda-lime glass is made by melting 72 percent silica (silicon dioxide), 15 percent sodium dioxide, and 9 percent calcium oxide. The remaining four percent consists of minor ingredients that give the glass particular properties. This glass is inexpensive, easy to shape, and reasonably strong. In the United States alone, each year sees the production of something on the order of 25 billion glass containers, or about 116 containers for every man, woman and child!

Glass is made in a variety of recipes. For example, if lead oxide is substituted for calcium oxide, the result is lead glass. This is a lustrous glass noted for its beauty and ease of working. Lead glass is particularly plastic when hot and is very strong and easy to work. This glass is much more expensive than soda-lime glass and is used for art objects, tableware, and the faces of TV tubes, because of its X-ray absorption characteristics.

Borosilicate glass is about 80 percent silica, 2 percent alumina (aluminum oxide), 4 percent alkalai, and nearly 13 percent boric acid. This glass was introduced by Corning Glass Co. under the name Pyrex®. This glass is about three times as heat and shock resistant as soda-lime glass and has a smaller temperature coefficient. It is used for ovenware, chemical plumbing, and telescope mirrors, where the extreme stability is valuable.

The most durable and sturdy glassware is made from a 96 percent silica recipe. A special borosilicate glass is prepared. Then some of the constituents are leached out, leaving the glass spongy. The remaining skeleton is fired at a very high temperature. It shrinks by as much as 35 percent in this process, which causes the pores to close. An extremely dense, transparent, nonporous glass is obtained. This glass is used principally for laboratory ware and is sold under the name Vycor® by Corning Glass Co. It is so temperature resistant that it can be heated red hot and plunged into ice water without cracking. It is also very strong and hard so that very thin walls suffice on test tubes and laboratory apparatus.

You can get a good idea of the vast variety of glasses that can be made from the fact that Corning alone has developed over 80,000 varieties of glass! Each recipe has some particular feature that makes it particularly useful in certain applications.

The early glassware made by the Egyptians and the Romans was always colored because of the accidental inclusion of certain elements in the glass. Glass is particularly sensitive to the inclusion of metal oxides. For example, one part of nickel oxide in 50,000 parts will tint the glass with a color ranging from bright yellow to purple, depending upon the type of glass. One part of cobalt in 10,000 parts will give an intense blue. The red glass used in car taillights is obtained with traces of gold, copper, or selenium oxides. You can see that the utmost care must be exercised in making color-free optical glass.

The degree of success achieved in this control is evident in the quality of the optical glasses now produced by some makers. A piece of good quality window glass seems perfectly colorless when you look through it. However, if the plate is viewed from its edge so that the light is traveling through about eight or 10 inches of the glass, it will invariably show a very strong color, generally green. A good quality plate glass will show less color, but a few feet will still be strongly colored. In contrast, a display at the Glass Museum in Corning, NY, has a bar of optical glass made for a submarine periscope which measures about three inches by four inches by 33 feet. A playing card is placed at the far end of this bar. The card can be read through 33 feet of glass without any noticeable distortion of the image shape or color. When you consider the fact that optical fibers are being offered with attenuations of only a few dB/km in selected bands, you can appreciate the care in preparation which such materials must require. If the Pacific Ocean were this clear, we could easily see the bottom of the Marianas Trench!

Most of the makers of optical fibers do not advertise the contents of the glass used in their fibers. However, it seems likely that most of the common fibers are a very high silica content glass, perhaps nearly pure silicon dioxide. In making stepped index fibers, the usual process is to take a tube of the lower index glass and shrink it onto a rod of the higher index glass. Technically, glass is a supercooled liquid since it has no crystalline structure. It also has no well defined melting point. As glass gets hotter, it simply decreases in viscosity and gets more and more pliable until at very high temperatures it can be poured like oil or syrup. This behavior is quite different from that of ice, which passes from a crystalline state to a liquid in a very small temperature range. Iron is plastic for only a small range and passes from a crystalline solid to a liquid in only about a 100°C span. Because of the wide plastic range, it is very easy to draw glass into very fine fibers. This can be done with a piece of laboratory glass tubing by heating the tube to incandescence over a gas flame. If opposite ends of the tube are pulled while the glass is plastic, the plastic portion will stretch nearly indefinitely. Interestingly, the round cross section and the hole in the tube will be retained even when it is drawn out to be finer than a human hair.

To draw glass fibers the original billet must be maintained at the correct temperature. Also, the drawing must proceed at a very precise and carefully determined rate. The exact procedures and techniques used for these operations are often trade secrets. Therefore, only practitioners normally know them.

CREATING GRADED INDEX FIBERS

The beginnings of the graded index fiber can be traced to about 1936 when R.K. Luneberg studied the properties of a peculiar type lens, which has since come to be known as a *Luneberg lens*. The Luneberg lens is unusual in that it is round like a marble. Parallel rays from *any* direction will form a real image on the surface of the lens opposite the direction of arrival. The Luneberg lens also has no distortion at all. The action of the Luneberg lens is obtained from the fact that the index of refraction varies as:

$$\eta = \sqrt{R_{MAX} - r^{2^l}} \qquad \text{Equation 11-1}$$

The index of refraction varies as the radius changes.

The ability to focus an image from any direction with no distortion is a very attractive feature. There would have been a great deal of interest in the Luneberg lens, except for one minor problem: Nobody had the vaguest notion of how to make one! During the 1950s, however, this changed. By this time a number of firms had developed techniques to make foam materials with arbitrarily adjusted dielectric constants. These foams could be molded into a series of shells, and a microwave Luneberg lens could be built up with a series of shells, each of which was only a fraction of a wavelength thick. The property of focusing energy in any direction with no distortion made the Luneberg lens very popular for a number of radar applications. The technique would not work at optical frequencies because the shells would have to be impossibly small and fragile.

The next major contribution came from a somewhat different source. With the invention of the transistor in the early 1950s there came a great technological impetus to develop the techniques necessary for transistor production. One of the first of these was the development of vacuum furnace techniques for producing ultrapure silicon crystals for the transistors themselves and for the controlled doping of the slices of silicon taken from the crystals. Also required were techniques for evaporating layers of silicon dioxide on the silicon surfaces for passivation. These techniques eventually made possible the production of graded index fibers.

If a glass rod is placed in a vacuum furnace and a gaseous dopant is introduced into the furnace, the dopant will penetrate into the glass rod that is held in the plastic state. In a plastic condition, glass becomes a conductor of electricity and can then be heated by currents induced by a coil energized with high frequencies. As the dopant diffuses, it will establish some profile, with the outer surfaces being the most heavily doped and the inner surfaces being the least heavily doped. If the dopant reduces the index of refraction slightly, the conditions are established for the construction of a graded index fiber. Once doped, the rod can be drawn out in a fiber-drawing operation similar to the one used with the stepped index material.

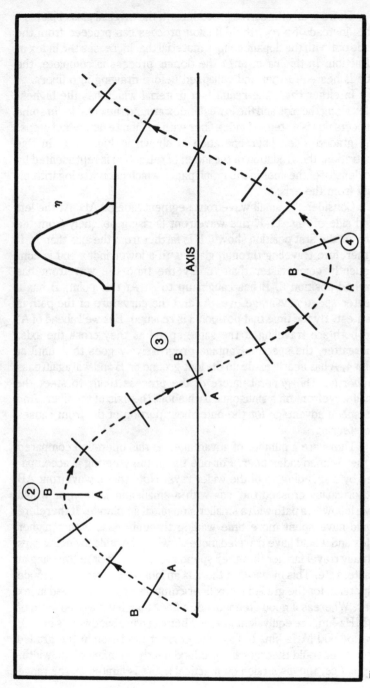

Fig. 11-1. The graded index fiber.

There is an alternative procedure. If the original glass billet is a tube, instead of a rod, the diffusion process can proceed from the inside out with the dopant being a material that increases the index of refraction. In this case, after the doping process is complete, the tube is heated further and collapsed before drawing into fibers.

In either case, the result is a material which has the highest index along the axis and the lowest index around the outside. In some respects this is a stepped index fiber with an infinite number of steps. The graded index fiber operation is shown in Fig. 11-1. In this illustration, the gradation of the index of refraction is represented by the density of the lines in the graph paper which decrease logarithmically from the axis.

Consider the small wavefront segment labeled AB. At the left hand side of Fig. 11-1, this wavefront is rising obliquely from the axis. At the first position shown, B is farther from the axis than A. It is therefore traveling through glass with a lower index and is consequently going faster than A. By the time the wavefront has reached position 2, B has caught up to A. At this point, B has a greater speed advantage over A, and the curvature of the path is greatest. By the time that position 3 is reached, B is well ahead of A, and both are traveling at the same speed as they cross the axis. Thereafter, the speed advantage progressively goes to A until at point 4, A has nearly made up the lost ground on B and will again pass B shortly. The general nature of the process tends to steer the small wavefront on a sinusoidal path about the axis of the fiber. And the speed advantage for the outermost portion of the front causes the steering.

There are a number of advantages to this operation compared to the stepped index fiber. For one thing, the steering is accompanied by a speeding up of the wider rays. Note that if wavefront AB had originally crossed the axis with a smaller inclination, it would have followed a path with a smaller sinusoidal amplitude. It therefore would have spent more time wading through glass with a higher index and would have traveled more slowly. The wide swinging rays actually travel farther, but they spend most of their time traveling at a faster rate. This means that there is an inherent advantage in mode dispersion for the graded index fiber compared to the stepped index fiber. Whereas a good stepped index fiber may have a bandwidth of 50 MHz-km, the equivalent graded fiber can have bandwidths of 200, 400, or 600 MHz-km. In fact, the correct gradation in the graded index fiber could theoretically produce nearly an infinite bandwidth. In practice, the dispersion on practical fibers is limited by the range of refractive indices and the control available in the diffusion process that dopes the index into the original billet. A number of manufacturers offer graded index fibers of the same external physical characteristics but with a spread in bandwidths and prices to match. This sounds very much as though a sorting process takes place in the fiber production, with the best product being sold at premium prices. This is not unusual in the integrated circuit business where the devices are manufactured in large batches, tested, and finally numbered and priced according to the test performance. If this is the case, it seems likely that continued volume production will provide increasing percentages of the premium grade product and substantial learning curve price reductions.

One characteristic of the graded index fiber is that for a given fiber diameter, the numerical aperture is generally smaller than it would be on a stepped index fiber. For a stepped index fiber with a loss on the order of 12 dB/km, the NA will tend to run between 0.2 and 0.35, whereas a graded index fiber with an attenuation between 5 and 10 dB/km of about the same physical size will tend to have a NA on the order of 0.16 to 0.2. The smaller NA is not an advantage because it makes it harder to launch power into the fiber.

The manufacturers are not at all standardized on the matter of measurement. For example, one current catalog states numerical aperture is measured in the far field with all modes excited on a 3-meter length of the fiber at the 95 percent power points. A second manufacturer measures NA at the 100 percent power points (which is questionable) on a 2-meter length of fiber. Some light is always scattered at nearly right angles, as shown in Fig. 11-2. Therefore, a 100 percent power measurement would always yield a NA of nearly 1. Two other catalogs simply give a value for NA with no reference to the power level at which it was measured or the length of cable used.

From the standpoint of the user, the NA measured at the 50 percent power points or the 10 percent power points would probably be more informative than any of the measurements quoted. Unfortunately, the NA obtained from these points would be smaller than the NA for a lower power level, and it would make the fiberguide seem harder to drive. The manufacturers therefore tend to practice a little "specmanship" by quoting the NA at lower levels to make the spec sheet more attractive.

One of the characteristics of the graded index fiber is that it tends to be generally lower in attenuation than the stepped index fiber. Whereas a very good stepped index fiber will seldom go below 10 dB/km, a number of manufacturers offer graded index fibers with

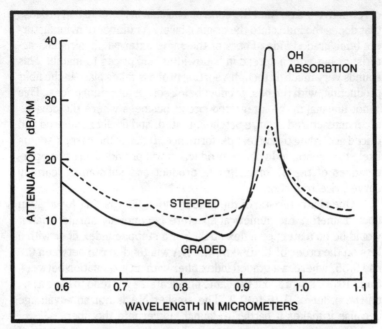

Fig. 11-2. Attenuation comparison of graded index and stepped index fibers.

losses all the way down to 3 dB/km. One manufacturer in particular offers a line of graded index fibers with attenuations measured at 900 nm of 8, 5, 4, and 3 dB/km. For each of the premium fibers, there is a subclass with a bandwidth of 200, 300, and 400 MHz-km. The fibers are physically identical and have the same NA of 0.2 so it seems likely that the offerings are the result of a product sorting. For these fibers, the NA is specified at the −10 dB (10 percent power) level with an unspecified length of fiber.

The actual reasons for the advantage shown by the graded index fiber in attenuation are not immediately obvious. There is no fundamental reason why a graded index fiber should have lower attenuation than a stepped index fiber. At least in theory, they could be made of the exact same kind of glass with the same amount of energy absorption. However, as a practical matter, in the stepped index fiber any irregularity at the interface between the core and the cladding will contribute to mode conversion and some reflection loss. In the graded index fiber, such an interface does not exist. Therefore, the graded index fiber has no real counterpart for this loss mechanism in the stepped index fiber. Obviously, both types have a certain amount of loss from microscopic bubbles and actual attentuation due to dielectric hysterisis, etc. There are also attenuations caused by impurity absorption. However, the prime culprit seems to be the interface dissipation of the stepped index fiber. It is entirely possible that improved processing techniques could put the stepped index fiber on the same footing as graded index fiber in terms of attenuation.

ATTENUATION CURVES

One area in which the graded index fiber is notably inferior to the stepped index fiber is the area of absorption band attenuation. Figure 11-2 is a repeat of the absorption curve of Fig. 10-2, with the absorption curve for a graded index fiber from the same manufacturer added. The attentuation for the graded index fiber is lower at most points; however, the hydroxyl absorption peak is much higher. In the fibers from other manufacturers, the peak at about 770 nm is also much higher for the graded fibers. Depending upon the application, this may or may not be a disadvantage. For most practical communication links, the wavelength can be selected to fall between the absorption peaks. The peaks may then have no effect whatever upon the design of the link. On the other hand, for broadband optical applications, such as viewing plates and fiberscopes, the absorption peaks are a distinct drawback if a full range of color is to be handled.

It seems likely that the increased absorption peaks in the graded index fibers are an artifact of the more extensive handling of the original billets during the diffusion process. If this is the case, it is possible that this disadvantage will eventually disappear as the diffusion technique improves. As is the case in the manufacture of the silicon devices used in solid-state electronics, the purity of the product is of tantamount importance in the production of usable devices. In the late 1950s, the silicon ingots were purified by a zone-melting process in which the ingot was suspended inside an evacuated quartz tube. The zone was heated with an induction heating coil that was a fraction of the length of the ingot. This molten zone slowly moved from one end of the ingot to the other so that the impurities could be swept along with it. Most of these impurities would end up at one end of the ingot. This impure end was sawed off and discarded.

The ingots were then about 1 to 1.5 inches in diameter, and thin slices were sawed off the ingot and polished. The N or P layer was then deposited upon the wafer, and the transistor electrodes were produced by an etching process. With this technique, the surface of the small wafer was covered with a pattern of transistors in a tiny checkerboard. The wafer was next scribed and broken up into "dice," with each die containing a transistor. The transistors were individually tested. Usually, approximately half of them were useless because of impurity defects in the wafer. The fraction of good parts was known as the *yield*.

These processes essentially demanded a high degree of automation. The original drawing of the silicon crystal and the zone refining process required levels of steadiness and control far beyond human capability. Under automatic control, however, it was possible to fine tune the process. Bit by bit, the yields began to increase. It eventually became possible to graduate to 2-inch and 3-inch wafers. Today, some manufacturers are using 4-inch wafers, and the distribution of imperfections is large enough so that reasonable yields are obtained with silicon chips as large as 150 mils on a side. Without this reduction in imperfection density, the microprocessor and semiconductor memory would be a financial impossibility because only a few of the 800 or more dies would be free of imperfections.

In many respects the problems involved in the production of high-quality fiberguides are similar. The original material must be highly purified. It then must be precisely doped to obtain the desired properties. It must then be drawn under extremely tight temperature and rate control. It seems likely that the fiber producing processes will benefit from the same factors that have benefited the transistor manufacturing process. The premium product of today will probably become the standard or low-grade product of tomorrow.

MEASUREMENT OF PASS BANDS

For the fiberoptic guide, some measurement technique is required to measure the attenuation of lengths of fiber over a wide range of wavelengths. This measurement is generally made with a device known as a *monochrometer*. This monochrometer is in many ways similar to the monochrometers used for measurement of light transmission of chemicals. However, it has certain special requirements when used for fiberguide measurements. Some of these features are shown in Fig. 11-3.

In the device shown in Fig. 11-3, the light from an intense incandescent lamp or a small arc is concentrated by a condensing lens upon the jaws of a slit. The slit has sharp edges like a razor blade. The light from the slit, or a portion of it, illuminates a ruled, concave, diffraction grating mirror. If the rest of the obstructions were not in the way, an image of the slit would tend to form along the arc P, P', and P'', with the shortest wavelengths imaging at P and the longest wavelengths at P''. The portion of the spectrum which strikes the fiberguide under test can be controlled by rotating the mirror to sweep the spectrum past the fiber.

The nature of a rotatable diffraction grating mirror is to have a relatively long focal length. Therefore, the angle between the converging light rays from the mirror will typically be much smaller than the acceptance cone angle of the fiber. The NA lens is therefore inserted to make the radiation converge more rapidly, and Θ_2 should be somewhat larger than the acceptance angle of the fiber. This more rapid convergence reduces the resolution of the monochrometer by the reciprocal of the ratio of the focal length with and

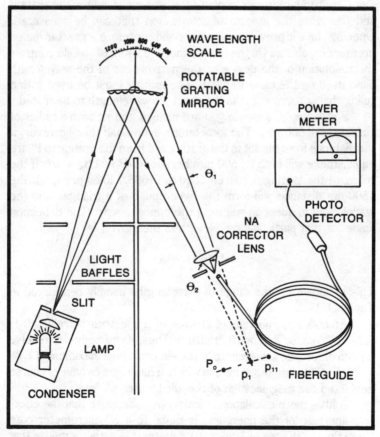

Fig. 11-3. The fiberguide insertion loss monochrometer.

Table 11-1. Wavelength Versus Deflection Angle.

wavelength	angle
400 nm	15.466°
600	23.578
800	32.230
1000	41.810
1200	53.130

without the lens. Consequently, this lens should be small and have the shortest focal length possible, consistent with the mechanical mounting requirements.

The width of the slit controls the amount of light in the system and therefore the amount of attenuation that can be measured. Opening the slit provides more light and therefore a greater measurement capability. On the other hand, the width of the slit controls the resolution of the device or the narrowness of the wavelength band used on the measurement. The grating must be used with a ruling that is narrower than the shortest wavelength to be tested.

For example, suppose you are using a grating with a radius of curvature of 500 mm. The focal length is one-half this; however, if the distance from the slit to the grating and from the grating to P' are equal, these will both be 500 mm because $1/P + 1/Q = i/f$. If the ruling of the grating is spaced 1500 nm (666.7 lines per mm), the 1500-nm distance will form the hypotenuse of a triangle, and the wavelength in question will form the opposite side. The deflection angle for that particular wavelength is then given by:

$$d \sin \Theta = \lambda$$

This is shown for the range of wavelengths usually considered in Table 11-1.

At 800 nm, the rate of change of angle with wavelength is 0.04516°/nm, or 7.88×10^{-4} rad/nm. Thus, if you wished to resolve a 1-nm difference in wavelength, the slit could be a maximum of $7.88 \times 10^{-4} \times 500$ mm, 0.394 mm, in width. These figures have of course neglected the magnification of the slit by the NA lens.

A little more calculation is instructive. Suppose that the effective aperture of the fiberguide is equal to a 50-micrometer core diameter, or an area of 1.96×10^{-9} meters^2. Further suppose that the effective area of the fiber projected at the slit is the same and that the source lamp has a power rating of 150W and that the effective lamp power at the slit is the same as if the lamp were to uniformly illuminate a sphere of 6-cm radius. The lamp power density at the slit would then be 3315 watts per square meter. This sounds like a great deal of power, but the tiny area of the image of the fiber would capture only 6.51×10^{-6} watts! And the bad news is not over yet.

If the mirror were not to disperse the light, the previous figures would give the amount of power entering the fiber. However, in practice the image would be smeared out over the arc from 15.46 degrees at 400 nm to 53.13 degrees at 1200 nm. With an image distance of 500 mm, this would be an arc length equal to 0.1645 meters and a height equal to the 50 micrometers of the fiber. For any one color, assuming the light was evenly distributed, the fiber would therefore pick up only one part in 4189 parts of the total light. And the light of any color launched in the fiber would be 1.55×10^{-9} watts. This is a very small fraction of our original 150 watts.

This tells you something about the design of these instruments. In order to have the light at a useful level, the instrument must be built faster; that is, with smaller dispersion and less resolution. A commercial instrument, such as the Times Wire and Cable model 125, has about one-tenth of the resolution of the example. The light output varies as the reciprocal of the square of resolution, so this unit would supply about 100 times as much light to the cable.

The instrument can be made faster by shortening the focal length of the grating. This is done either by reducing the radius of curvature of the grating itself or by including lenses. The reduced resolution of the instrument will reflect itself mainly in the loss of the fine detail of the absorption peaks. When the measurement is performed on two instruments with different resolution, the higher resolution machine will generally show higher peaks on the absorption lines. In general, the pass band windows are sufficiently broad so that these will be the same in both instruments. For example, the absorption curve for the graded index fiber shown in Fig. 11-2 was probably made with an instrument that had a resolution of about 10 to 15 nm. An instrument with a 1-nm resolution might have showed an OH peak that was not so rounded on top and had a height of 50 dB/km. Generally, the achievement of a very high resolution is more important to someone trying to diagnose the source of the losses in the cable and fine tune the processes to eliminate them. For the person interested only in using the cable, the exact nature and height of the absorption peaks is mainly of academic interest. The cable is seldom used in the vicinity of these peaks anyway.

There is yet another problem involved in developing a suitable test set for fibercables. The output curve of the light source and the curve of response for the detector both tend to be functions of frequency, and the amount of light coupled into the fiber is a function of the active size of the fiber and some other factors as well.

One way in which most of these factors can be eliminated is to use the test set to measure some length of fiber and then shorten the fiber by a known amount and repeat the measurement. The difference between the two curves and the known difference in length is used to furnish an absolute curve of attenuation versus wavelength. The same coupling losses and source and detector laws are present in both measurements and will not appear in the difference readings. This type of measurement is well suited to microprocessor control. The processor can drive the wavelength scale through a stepper motor—and remember the output readings. The cable can then be shortened and the readings repeated. If the difference in length is entered, the processor can then determine the differences between the readings and perform the manipulation with the known length so that the output is plotted directly in terms of dB/km.

No one yet knows whether there is some absolute bottom limit to the expected absorption in the band pass windows of good fibers. Measured in terms of the absorption per wavelength of signal path, these losses in premium cables are already many orders of magnitude below the most efficient microwave waveguide or electrical transmission line. The losses in the existing premium cables are set largely by the presence of submicroscopic bubbles and imperfections and the inclusion of impurity elements. Since both of these are subject to improved process control, the possibility of further advances remains.

12

Losses in Coupling

One of the reasons the monochrometer was covered in Chapter 11 is that it serves as an edifying introduction to the subject of the losses involved in the coupling of energy into a fiberguide. This discussion will help you to understand some of the reasons for the popularity of certain types of sources and detectors in fiberoptic work.

SOURCE SIZE MATCHING

The very fact which is responsible for some of the popularity of the fiberguide is also the source of some of the major problems involved in engineering a suitable fiberoptic system; that is, the fibers are so confounded small. This blessing makes using fibers in place of copper wire or waveguides very attractive, but it also gives design engineers fits in the area of hardware.

To begin, consider the matter of *unintercepted illumination* and *gain*. For many sources, such as a spark and a lamp filament, there is no particular preferred direction of radiation; the activity tends to shed energy about equally in all directions. There are always areas in a practical source which are not illuminated due to the electrodes for the spark or the lead-in wires and support for the lamp filament. However, the nature of the spark and the filament is to radiate energy equally in all directions. In a device of practical size, some minimum distance from the practical source must be maintained. For example, in the lamp filament, it is not physically possible to have the fiber actually touch the filament because the filament is so hot that it would melt the fiber. A similar factor applies to the electric arc. In the case of the incandescent lamp, the fiber must usually be outside the lamp envelope. Inverse square law spreading applies, and the fiber will be able to intercept only a fraction of the energy produced at the source. That fraction will be equal to the ratio of the area of the system pumping energy into the fiber compared to the area of the sphere or the equivalent sphere illuminated by the lamp.

To slightly amplify this point, you might consider another example. For the monochrometer, a projection lamp with an output of 150 watts was used over the region from 400 to 1200 nm. This lamp would probably have required a 400-watt input electrically to produce that visible output in that range, since the predominant portion of the output is in heat at much longer wavelengths than 1200 nm. Because of the cooling requirements and the size of the optical condensing system, the radius of the system was taken as 6 cm. You probably wonder whether a smaller lamp with the correspondingly smaller radius might not work better. Suppose, for example, that you were to try an automobile taillight. For such a lamp with an input rating of 12 volts × 3 amperes, the envelope radius is about 1.5 cm. The area of the sphere is thus 2.83×10^{-3} square meters. This would be flooded by one-fourth of the electrical input, or 9 watts, to give 3183 watts per square meter. This is scarcely an improvement over the projection lamp/condenser system, which calculated to 3315 W/m^2.

This brings us head-to-head with a limitation of the design of incandescent lamps. For a soda-lime envelope, a visible light power density in excess of about 3000 W/m^2 brings the glass dangerously close to softening unless good natural convection cooling is available. Any attempt to run the lamp higher tends to require forced air cooling. If this is accepted as an upper limit, a fiber area like 2×10^{-9} m^2 will only accept about 6 microwatts from the source in light of all colors.

CONDENSING SYSTEMS

Having bumped your head against the power density limitation of incandescent lamps, it seems logical to ask whether there might not be some optical reducing system that could increase the amount of power incident upon the end of the fiber. The lamp filament is obviously larger than the fiber itself; you are probably inclined to wonder whether something like a backward microscope might reduce or condense the size of the image to the point where it is no longer than the end of the fiber. Perhaps this would help the situation.

200

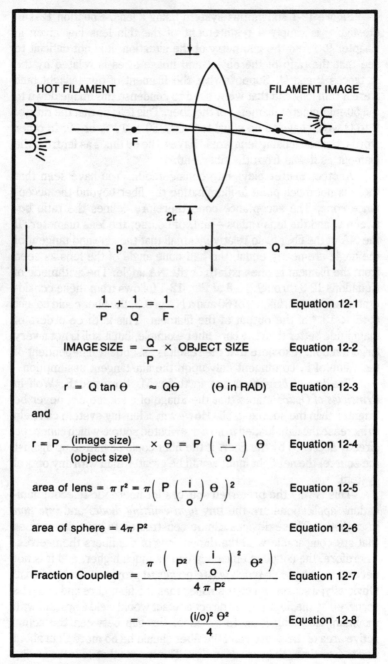

$$\frac{1}{P} + \frac{1}{Q} = \frac{1}{F}$$ Equation 12-1

$$\text{IMAGE SIZE} = \frac{Q}{P} \times \text{OBJECT SIZE}$$ Equation 12-2

$$r = Q \tan \theta \approx Q\theta \quad (\theta \text{ in RAD})$$ Equation 12-3

and

$$r = P\frac{\text{(image size)}}{\text{(object size)}} \times \theta = P\left(\frac{i}{o}\right)\theta$$ Equation 12-4

$$\text{area of lens} = \pi r^2 = \pi\left(P\left(\frac{i}{o}\right)\theta\right)^2$$ Equation 12-5

$$\text{area of sphere} = 4\pi P^2$$ Equation 12-6

$$\text{Fraction Coupled} = \frac{\pi\left(P^2\left(\frac{i}{o}\right)^2\theta^2\right)}{4\pi P^2}$$ Equation 12-7

$$= \frac{(i/o)^2\,\theta^2}{4}$$ Equation 12-8

Fig. 12-1. The optical condensing system.

Figure 12-1 shows this system using a lens. Equation 12-1 of the figure is simply a restatement of the thin lens law given in Chapter 9. From the geometry of the situation, it is not difficult to see that the ratio of the object and image sizes is related by the distances P and Q. Suppose that the filament of our taillight bulb were 3-mm long and that we wished to condense this image down to the 50-micrometer diameter of the fiber. This tells us that the ratio of P to Q would have to be equal to $0.003/50 \times 10^{-6}$, or 60. In other words, the condensing lens would have to be 60 times as far from the filament as it was from the fiber end.

Another matter plays into consideration. You have seen that there is not much point in illuminating the fiber beyond the acceptance cone. The acceptance cone therefore defines the ratio between Q and the lens radius—and, of course, the lens diameter. If the NA of the fiber is so relatively small that the sine and tangent of the angle are nearly equal, the half cone angle of the lens as seen from the filament is one-sixtieth of the NA angle. The arithmetic of Equations 12-3 through 12-8 of Fig. 12-1 follows from these considerations. For a shrinkage of 60 and a NA of 0.3, the fiber could collect 6.45×10^{-6} of the output of the filament. This is three orders of magnitude better than the bare fiber example, but it still is not a very large fraction. Also note that the result is essentially independent of the value of P, contingent only upon the sin/tangent assumption.

The *law of brightness* set forth by M. Born and E. Wolf in *Principles of Optics* states that the image of a source can never be brighter than the source itself. However, a lensing system can help to increase the light loaded from a distributed source which cannot be directly accessed by the fiber. If the fiber could be butted up against the source, the net light input would be greater than with any optical system.

This is why the preferred sources for fiberoptic telecommunications applications are the tiny *light-emitting diodes* and *injection laser diodes*. These devices can be constructed to have dimensions that are comparable with the dimensions of the fibers themselves. Therefore, the coupling efficiency can be much higher, and it is not necessary to have so much waste power to provide a usable signal. Obviously, a source which is smaller than the fiber core and could be accessed by the fiber core in direct contact would yield a system with minimum launching loss. In practice, the gap between the actual active area of the source and the fiber should be no more than about two to four times the core diameter. This means that sources having a lens are at a distinct disadvantage since the lens must usually be placed 1 to 2 mm from the active part of the source, whereas the limit would call for a spacing of 100 to 200 micrometers—about one-tenth this distance. To this end, many manufacturers will supply the sources with a factory installed "pigtail" of fiberguide that has already been bonded to the active area of the emitter.

LAMBERT'S LAW AND DIRECTIVITY

If a piece of flat sheet metal is heated to incandescence and an instrument is then used to measure the amount of light given off in any direction, the amount of light will vary as the cosine of the angle off of the normal to the flat. The light will not collimate into a sharp beam because the radiation given off from any one point on the surface is completely independent of the light being given off from any other portion. The thermally induced radiation is said to be *incoherent,* and it is smeared out over a hemisphere. Viewed from the edges—that is, in the geometric plane which contains the surface of the flat specimen—the light output will be zero. This cosine relationship between the distributed power and the angle from the normal is termed *Lambert's law.* A piece of ground glass illuminated from behind, one of the cold discharge nightlights sold in dimestores, or the light reflected from a flat sheet of white paper will all display a lambertian distribution.

Compared to a point source of light, however, such sources have a characteristic called *gain* or *directivity.* As noted in Chapter 3, the point source distributes its energy equally in all directions. By comparison, the Lambert's law distribution has a very definite maximum along the normal to the plane surface and it rapidly drops to zero elsewhere. For this reason it is considered *directive:* It does not need to have as much power all told to generate an equal power in the direction along the axis.

For practical, solid-state sources in fiberoptic communications work, the distribution of the energy is often even sharper than the cosine law. A study of the catalog sheets for various sources will often yield a radiation pattern similar to that in Fig. 10-6. Such patterns can be measured by mounting the device in a fixture arrangement, as shown in Fig. 10-5, and measuring the amount of light output as a function of angle.

It is frequently desirable to be able to calculate the directivity of the source and determine the total amount of power from the power density along the strongest direction. This can be done by adding or integrating the power readings taken in all directions from the source. However, an analytical expression is often not available to

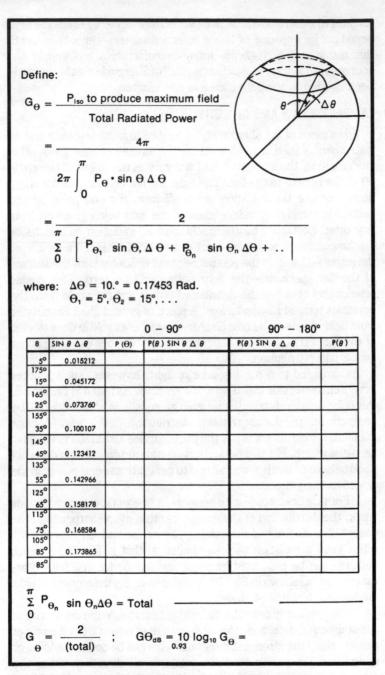

Define:

$$G_\theta = \frac{P_{iso} \text{ to produce maximum field}}{\text{Total Radiated Power}}$$

$$= \frac{4\pi}{2\pi \int_0^\pi P_\theta \cdot \sin\theta \, \Delta\theta}$$

$$= \frac{2}{\frac{\pi}{\sum_0} \left[P_{\theta_1} \sin\theta_1 \Delta\theta + P_{\theta_n} \sin\theta_n \Delta\theta + .. \right]}$$

where: $\Delta\theta = 10.° = 0.17453$ Rad.
$\theta_1 = 5°, \theta_2 = 15°, ...$

θ	SIN θ Δ θ	P (θ)	P(θ) SIN θ Δ θ	P(θ) SIN θ Δ θ	P(θ)
			0 – 90°	90° – 180°	
5°	0.015212				
175°					
15°	0.045172				
165°					
25°	0.073760				
155°					
35°	0.100107				
145°					
45°	0.123412				
135°					
55°	0.142966				
125°					
65°	0.158178				
115°					
75°	0.168584				
105°					
85°	0.173865				
85°					

$$\sum_0^\pi P_{\theta_n} \sin\theta_n \Delta\theta = \text{Total}$$ ———— ————

$$G_\theta = \frac{2}{\text{(total)}} \quad ; \quad G\theta_{dB} = 10 \log_{10} G_\theta = $$
0.93

Fig. 12-2. Directivity of axially symmetrical radiation patterns.

204

cover the measured response of the source. The table in Fig. 12-2 was developed for this purpose.

As seen in Fig. 12-2, the calculation operates on the assumption that the radiation from the source is symmetrical about some axis of the system. All that is required is to add all of the powers incident upon each of the bands of the imaginary measuring sphere. If you assume the sphere has a radius of unity, the area of any band is simply $\sin\Theta\,\Delta\Theta$. The value of $P(\Theta)$ is simply taken from the measured data, written in the appropriate column and then multiplied by

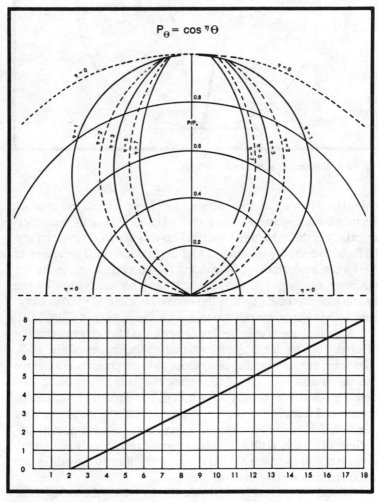

Fig. 12-3. Directivity estimation.

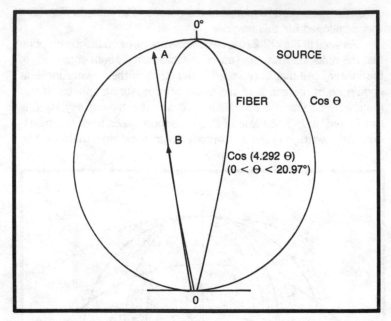

Fig. 12-4. Numerical aperture mismatch.

$\sin\Theta\Delta\Theta$. When all of the entries have been made, the product columns are added and the sum is divided into 2 to yield the numerical directivity of the measured pattern. This may also be converted into decibels. Directivity is commonly given in decibels. The values of $P(\Theta)$ must be normalized by dividing the absolute readings by the maximum reading. The readings may also be normalized by the measurement process so that the maximum value of $P(\Theta)$ is unity, and all other values are relative to it, as in the pattern of Fig. 10-6.

A second and somewhat faster technique for determining the directivity of a source is by estimation from the curves of Fig. 12-3. These curves represent the $\cos\Theta$ function to various powers and are thus algebraically solvable. The measured data is simply fitted over these curves and the closest curve noted. The directivity can then be interpolated from the graph. For example, a curve measured on a source which neatly fits the n=3 curve would have a numeric directivity of 8, or 9.03 dB. This source would require only one-eighth as much power to create a given signal strength along the axis as one which radiates equally in all directions.

Directivity is usually distinguished from gain by the losses in the system. In a light-emitting diode, for example, some of the electrical input energy is converted into heat and into wavelengths that are not usable. The actual gain would be the usable light output density at the peak of the radiation pattern, divided by the electrical power input that would be required to illuminate a sphere at the same level with a lossless source. In practice, the directivity of the source, minus the power losses for all causes, is the *gain* of the source.

NUMERICAL APERTURE LOSS

When the attempt is made to couple a source to a fiber, it is seldom that the numerical or numeric aperture (NA) of the two will match. Even when they do, a certain amount of NA loss exists. Consider the curves of Fig. 12-4. The rounder curve is a cosine or Lambert's law radiation pattern, which is typical of a LED source. A planar source, such as a LED which is not coherent, will typically produce a pattern such as this one in a measurement. The pattern is for the source alone, with the assumption that the fiber can be butted up against the source. If the source were equipped with a lens, the pattern would be much narrower.

The narrower inner curve is the curve that was fitted onto the fiber NA measurement of Fig. 10-2. This is representative of the acceptance response of the fiber. If you presume that the source is smaller than the fiber core, at any direction the amount of energy launched into the fiber will be proportional to the product OA × OB. Why? The source is launching less than the peak energy in this direction, and the fiber does not accept energy as efficiently in this direction. To see the net effect of this mismatch, refer to the gain computation of Fig. 12-5.

For this computation, the usage of the table in Fig. 12-5 has been slightly altered. Rather than use the right hand columns for the angles between 90 degrees and 180 degress, this portion is used for calculating the power distributed by the source. The left hand column is used for calculating the product of the power distributed by the source and the power accepted by the fiber. A substantial loss is caused by the fact that the LED source is throwing most of its energy in directions in which the fiber cannot accept it; in fact, 84.7 percent of the energy from the source is launched at angles where the fiber cannot accept it at all.

A second point must be kept in mind. Suppose that both the source and the fiber had identical patterns of P_θ = cosine (4.292 Θ). The actual amount of energy launched would be a function of the square of the pattern function, whereas the energy furnished would be only a function of the pattern. A calculation similar to that of Fig.

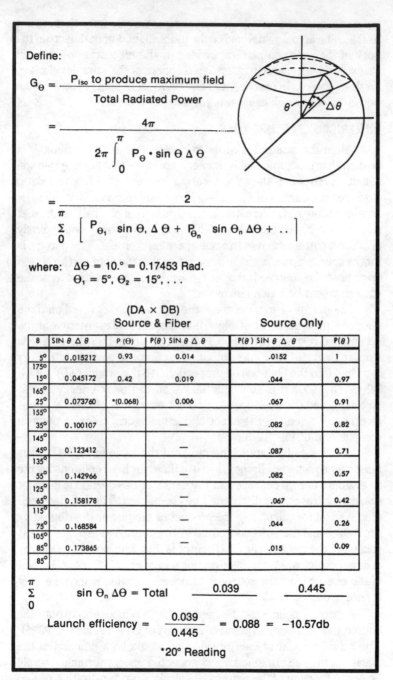

Fig. 12-5. Numerical aperture mismatch calculation.

208

12-5 would reveal that for this case, the launched power would represent only 0.612 of the generated power. A net loss of -2.129 dB would still exist, despite the source and fiber patterns being identical. Only when the acceptance pattern of the fiber is much wider than the launch pattern of the source can high launching efficiencies be obtained.

The astute reader might argue that the pattern used was measured on a length of fiber longer than the coupling length and that the pattern to be measured on a fiberguide shorter than the coupling length would be wider and therefore present less in the way of losses. Furthermore, the argument set forth in Chapter 10 indicated that the narrowing of the acceptance pattern was attributed, at least in part, to mode coupling. This implies that some of the energy originally launched into the fiber at wider angles had been refocused toward the central modes. This is probably true, and the calculation thus performed is perhaps a bit on the pessimistic side. Measurements have demonstrated too much scatter to determine whether the wider modes are efficiently converted or simply dissipated. If there is an efficient conversion, the calculation is pessimistic; however, if the wider modes are not efficiently converted, the calculation will turn out reasonably accurate. In either case, if the acceptance pattern is measured with the type and length of fiberguide to eventually be used, the errors are not great.

Before departing from this subject, note that the amount of energy in the various annuli peaks at about 45 degrees for the Lambert's law source. The area in that annulus has grown faster than the power from the source has fallen.

IMPEDANCE MISMATCH

For typical indices of fiber and air, something on the order of four percent of the energy is lost on reflection. This fraction is negligible at the source, although it is not negligible at joints in the fiber along the line. However, you can neglect this loss completely in the calculation of NA loss because it is so relatively small.

LENSING THE FIBER END

It is relatively obvious from the foregoing discussion that if the fiber is larger than the source, the chief culprit in the matter of coupling is the NA loss. Would it be possible to do something to increase the NA of the fiber? Surprisingly, there is something to be done. Although it might seem physically impossible at first to do so, it is possible, and for that matter not too difficult to place a convex lens on the end of the fiber. It turns out that the end of a stripped fiber, when heated to its running point, will tend to draw into a spherical shape under the influence of surface tension. If the end of the fiber is held securely in place and only the very end is heated to incandescence with a small oxyacetylene tip (No. 2 or No. 1, with a blue cone no more than ⅛ inch), the fiber end will tend to draw into a nearly spherical shape. This takes a bit of practice, though. It can also be done with a single-bottle gas torch such as liquid propane or Mapp® gas; however, there is a net advantage to having the hotter flame since the end can be softened before much of the fiber is heated. A miniature oxyhydrogen torch of the type used for instrument work would probably be better still. Too much heating will produce a ball on the end of the fiber, and letting the flame play up on the fiber will produce a crinkled length.

Figure 12-6 shows a good lens. When properly done, such a lens will approximately double the width of the radiation pattern. Experiments at a number of laboratories have indicated that reductions in NA losses on the order of 3 dB are practically attainable. However, the formula for the focal length of the lens makes a number of assumptions whose validity decreases rapidly with departure from the axis of the system. This form of lens is severely plagued with spherical aberration and astigmatism. Therefore, the precise definition of a focal length is a bit questionable because the focus is rather poor. For a radius of 100 micrometers and an index of 1.48, Equation 12-1 of Fig. 12-6 gives a focal length of 308 micrometers. However, paraxial rays from 50 μm off axis would focus at 290 μm. For 25 μm off axis, the focus is at 302 μm.

NEAR-FIELD EFFECTS

One of the sources of error in the preceding arguments and techniques is that the radiation pattern measured for the end of the fiber is actually a far-field measurement, whereas the actual measurement of the coupling should really take place in the near field. To understand the difference between these, if you examine again the wave interference expression of Chapter 6, you will note that we assumed that the measurement point was sufficiently far away so that the rays from the individual sources could be considered parallel. The measurement apparatus shown in Fig. 10-5 also employed such a long distance that the parallelish condition was approximately satisfied. On the other hand, when you consider placing the fiber very close to, or even in physical contact, with the source, these conditions are obviously not satisfied.

Figure 12-7 shows that the beam collimation does not take place immediately in front of the lens or the aperture of a collimating device. In the figure, a dimensionless point source of light is placed at the focus of a lens. The lens is assumed to be essentially perfect so that the light rays are rendered parallel or collimated. Also, the beam spreading measured at some distance from the system will be limited only to the diffraction effects. At some small distance from the lens, the energy to be measured by passing a probe across the beam is not a very well defined affair. The interference of the waves from the various portions of the lens do not add in phase, and the waveform is somewhat jumbled. The pattern in space for the transverse cut, however, is still shaped pretty much like the lens.

At a somewhat greater distance, the interference effects are more evident, and a series of ripples has formed on the transverse cut. A well defined beam is still not available, though. At a still further distance, the phase addition has begun to smooth out the energy distribution. This process continues gradually until at the distance known as the *far-field limit,* the pattern of the antenna can be defined entirely in terms of angles and no further changes will take place. All of the polar diagrams measured at any greater range will be identical and will resemble the polar diagram shown in the figure.

The selection of a range of $2d^2$/wavelength for the far field limit is somewhat arbitrary. This is simply the point where the curvature

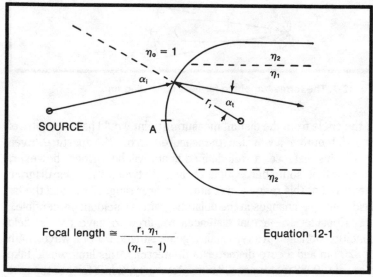

Focal length $\cong \dfrac{r_1 \eta_1}{(\eta_1 - 1)}$ Equation 12-1

Fig. 12-6. Lensing the fiber end to increase effective numerical aperture.

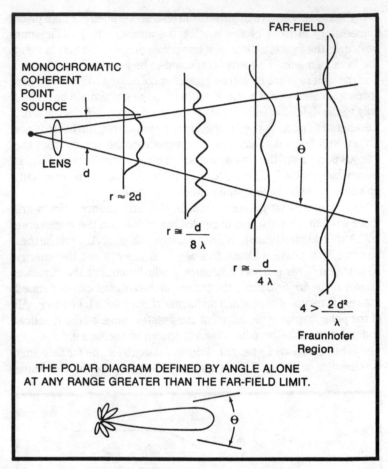

Fig. 12-7. The formation of the far-field polar diagram.

of the circle from the distant measuring point would have a sagitta of one-sixteenth of a wavelength measured across the aperture. Even at half this distance, a well-defined beam will be formed; however, there will be some perceptible changes between the polar diagram measured at this range and at much greater ranges. Beyond the far field limit, the changes in the polar diagram are seldom perceptible.

Consider the actual distances required to have the far-field condition obtained in a typical fiberguide situation. For a wavelength of 820 nm and a core diameter of 60 microns, this limit would take place at 0.0088 meters or 8.8 mm. Obviously, this is farther than the 120-μm to 240 μm spacing given earlier as a maximum for the fibers.

On the other hand, if you consider the beamwidth being generated in the polar diagram, you obtain a somewhat different picture. It was noted in Chapter 6 that for a diffraction limited aperture of circular cross section, the full width of the beam is given by 1.14 radians, times the wavelength, divided by the aperture diameter. For the previous fiber, this would work out to 0.0156 radians or 0.894 degrees. This is much less than the measured radiation pattern for the fiber. The obvious point is that the fiber is viewing an extended source and will therefore output an extended image. In other words, the polar diagram measured for the fiber is probably the superposition of each of the mode patterns.

The use of the far-field pattern in calculating the NA coupling loss certainly has some errors due to this factor. However, it seems as a practical matter that the calculation is reasonably consistent with experimentally derived results. This is probably due to some portion of the beam or pattern formation taking place at points other than the immediate face of the fiber. Therefore, the far-field approximations are more reasonable.

OVERLAP LOSSES

When two fibers do not precisely overlap and perhaps have different diameters, a loss will exist. Two fiber cores are shown in Fig. 12-8. In this general case, the diameters of the fibers are different and the two axes do not coincide. The overlap area can be broken down into two segments of circles. The efficiency is approximately equal to the common area, divided by the area of the source circle. The formulas in the figure show the mechanism for calculating this overlap area. Obviously, if one fiber core is larger, there will be some range where the loss is minimal and constant because the circle for the smaller core lies entirely within the circle of the larger core. This loss can theoretically be zero if the source core is the smaller.

In an actual situation this consideration is a bit oversimplified because the simple area relationship assumes that the power distribution in the cores is constant. Actually, this is not so. The power is denser near the center after the signal has traveled any significant distance down the fiberguide.

ANGULAR MISALIGNMENT LOSSES

If the fibers actually touch at one corner but have some angular misalignment, some of the energy emerging from the source end of the fiber will strike the second fiber at angles that are beyond the

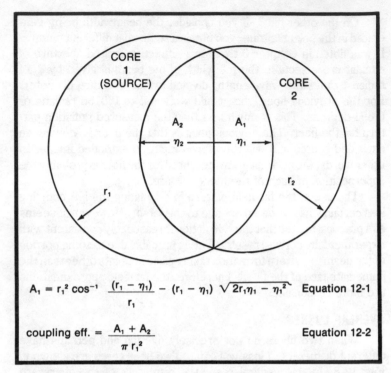

$$A_1 = r_1^2 \cos^{-1} \frac{(r_1 - \eta_1)}{r_1} - (r_1 - \eta_1) \sqrt{2r_1\eta_1 - \eta_1^2} \quad \text{Equation 12-1}$$

$$\text{coupling eff.} = \frac{A_1 + A_2}{\pi r_1^2} \quad \text{Equation 12-2}$$

Fig. 12-8. Overlap losses.

acceptance angle. This particular form of error is much worse for fibers with a large NA than for fibers with a small NA. With an NA of 0.5, a misalignment of 3 degrees will produce an insertion loss of about 1 dB. A fiber with an NA of 0.2 would require a misalignment of nearly 10 degrees for the same coupling loss. Some of the measures taken to avoid this type of loss will be covered later.

END SEPARATION LOSSES

At a fiber splice, if the two ends do not but up against one another, there is an end separation loss similar to the ones described for the source/fiber coupling. This loss is also very sensitive to the NA of the fibers, because a large NA fiber will permit more rapid spreading of the energy. Consequently, more energy will miss the core of the other fiber. Typically, a fiber with an NA of 0.5 will show a loss of 3 dB for an end-to-end spacing of 0.3 fiber diameters. For an NA of 0.2, an end-to-end spacing of one-half a fiber diameter will produce a loss of about 1 dB.

214

NA AND FIBER SIZE CONSIDERATIONS

The small size and small numerical aperture of high-quality fibers tend to make these items very difficult to use with anything except a special pigtailed LED or other solid-state source that has relatively small dimensions. In certain cases, however, you'll want a short run fiberguide system capable of working with more extended sources. For these cases, the plastic fibers are often selected. They generally have a rather large NA, and they are usually fabricated in larger diameters than the low-loss glass. Where the system length is measured in terms of a few feet, the relatively high losses for these fibers is usually not a major problem. Another advantage of plastic fibers is that they tend to have a reasonable transmission in the visible range. There are a number of applications where the fiber is only required to sample the light from some illuminating source and to present the information for detection by a human eye. One such application uses fiberguides to monitor all the lights of certain luxury automobiles. Each fiber separately samples the illumination in a parking light or a headlight and conveys this to the dashboard for inspection by the driver. A lamp that is not functioning is due to the lack of symmetry in the dashboard display.

In some communications applications, where the main object is to convey some limited information and you want to use a low-cost incandescent source and low-cost photodetector, the logical choice is sometimes one of the fiber bundles. These bundles used to transport telecommunications are generally referred to as *incoherent* since the fibers are not carefully oriented. The incoherent bundle cannot be used for image guiding because the image elements would be scrambled like the fibers. However, for simple light pickup from sources like an incandescent lamp, the large pickup area will collect many more times as much light from an extended source as a single fiber. An extended area detector used with an incoherent bundle will effectively sum the light input from all of the fibers. The popularity of incoherent bundles will probably diminish as high-quality pigtailed sources and detectors become less expensive.

13

The Quantum Theory of Light

At the outset of Chapter 4, it was mentioned that we must treat light in three distinct manners in any analysis of fiberoptic systems. In one set of circumstances, it would be necessary to consider light as a train of waves which had certain wavelengths and basically electrical properties. In the second case, we would treat light as though it were mathematically abstract rays which could be bent on refraction or reflection but otherwise had no physical properties. And in the final case, we would have to treat light as though it were made up of quanta of energy, little corpuscles similar to those that Newton had postulated, but having some rather different properties. The requirement for the quantum theory of light comes from the need to have some theory to describe the operation of photoemitters and photodetectors. Without the quantum explanation, there is no currently accepted mechanism for explaining the operation of these devices.

THE SPECTRUM

Our story of the development of the quantum theory must begin with the study of the spectrum since this provided the background for the theory. After Newton had demonstrated that white light could be broken up into a rainbow or spectrum of colors, a great deal of work followed in the field of what came to be known as *spectroscopy*. People studied the spectra of light for a variety of reasons, hoping to determine something of the properties of light itself and also some of the properties of certain chemicals. For example, it was known that table salt heated in a flame produced a brilliant yellow light, cadmium salts and lithium salts produced a red light, and copper salts produced a green light. It seemed likely that something could be learned about the nature of these things if a reasonable explanation of the source of the colors could be derived.

The basic structure of a *spectrograph* is shown in Fig. 13-1. The unit consists of a number of components. First, there is a light source which serves to illuminate the slit. The light that passes through the slit and the hole in the mask is collimated by lens A. In other words, the slit is at the focus of lens A so that A would tend to project an image of the slit focused at infinity. This ensures that the wavefronts reaching the prism are flat. The prism deviates the light and, because of the dispersion of the prism, the violet portion of the spectrum is deviated more than the red. However, for each color, the light emerges from the prism in flat wavefronts. It is still focused at

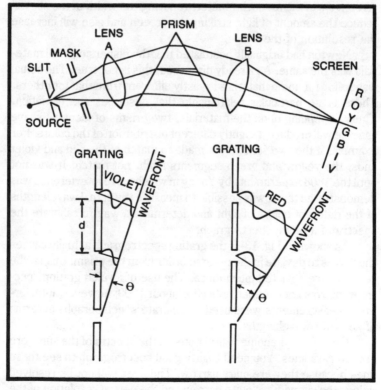

Fig. 13-1. The spectrograph. For a grating or interference spectrograph, the order of the colors is reversed to those of a prism.

217

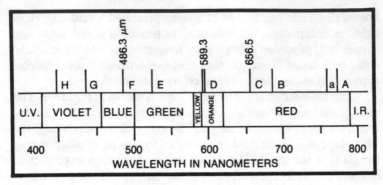

Fig. 13-2. The solar system and the Fraunhofer lines.

infinity. Lens B then focuses the parallel wavefronts onto the screen. The screen actually shows a series of images of the slit with the image for each color coming to a focus at a different place on the screen. As in the monochrometer, reducing the width of the slit will reduce the amount of light striking the screen and also will increase the resolution of the image.

Newton had originally postulated that the dispersion of all materials was the same. As already discussed, this is not true. The actual dispersion for some materials is vastly different from that for others. There is no particular relationship that describes the deviation. Thus, depending upon the materials, two prisms of identical shape and size will produce a slightly different distribution of the colors. For example, if the two spectra are made to match at the red and violet ends, the yellow and green segments might not match. It was not until the 1808 experiments by Young in which wave interference was demonstrated that it was possible to measure the actual wavelengths of the different colors of light and determine a way to calibrate the spectra of a prism spectrograph.

As shown in Fig. 13-1, the grating spectrograph actually causes the waves to deviate in the reverse order from the prism; that is, the red rays are deviated much more. The use of a ruled grating spectrograph was not common until after about 1900; however, interference measurements were used to calibrate spectrographs in terms of absolute wavelengths.

One of the phenomena first noted in the spectra of the sun were certain dark lines. You need a fairly good spectrograph to see these lines because they are quite narrow. The lines will not be resolved unless the slit width is quite narrow and the optical resolution of the spectrograph is quite good. By 1820, Joseph Von Fraunhofer had noted and identified some of the more prominent lines, though. Figure 13-2 shows a spectrum of the sun drawn to a linear scale with the colors marked out and the Fraunhofer lines shown. Fraunhofer was not able to resolve the two lines marked D at 589.3 μm. These lines were suspiciously close to the place in the spectrum occupied by the yellow light given off when salt was heated to incandescence in the flame of a bunsen burner. Fraunhofer also lent his name to the analysis which defines the near and far zones of a collimating lens or mirror, as noted in Fig. 12-7.

The next step in the drama came from a chemist, Robert Wilhelm Bunsen, and a physicist, Gustav Kirchhoff, working at Heidelberg, Germany. In 1855 Bunsen invented a gas burner or torch which produced an intensely hot flame using gaseous fuel. With this torch, Bunsen and Kirchhoff heated a number of chemical elements to incandescence and identified the spectra. They also investigated the spectra when elements were heated with electric arcs and electric sparks. They found among other things that the lower the pressure and the higher the energy imparted to the element, the brighter, sharper, and more numerous the lines became. They also found that each element had its own distinct signature. An element could be identified by the signature alone. The D line identified by Fraunhofer represented the element sodium. For the first time, one of the constituents of the sun had been identified! By creating an electrical discharge in a tube containing hydrogen at pressures well below atmospheric pressure, they were able to also identify the C, F, G and H lines as being part of the spectrum of hydrogen, thereby identifying a second element in the atmosphere of the sun.

One problem existed in the investigation: The Fraunhofer lines were dark, whereas the lines in their laboratory spectra were light. This problem was resolved when it was discovered that if the spectrograph was set up and the emission lines of an element were displayed, the bright spectrum was obtained; however, if the slit were then also illuminated with the brilliant white light of an electric arc, the lines appeared dark. Apparently the elements would absorb the same light that they gave off if the background were brighter than the emission.

The investigation of the spectra of the elements continued with tables of data being compiled on various elements. But no one was making much progress in identifying the cause of the spectra. Why would an atom like hydrogen give rise to discrete spectral lines that were very well defined in wavelength, rather than a broad continuum of radiation like an incandescent piece of iron or platinum?

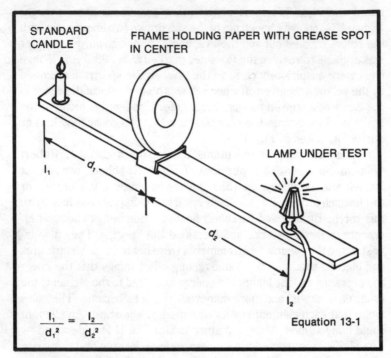

Fig. 13-3. The Bunsen photometer.

In the figure:

STANDARD CANDLE

FRAME HOLDING PAPER WITH GREASE SPOT IN CENTER

LAMP UNDER TEST

I_1

d_1

d_2

I_2

$$\frac{I_1}{d_1^2} = \frac{I_2}{d_2^2}$$

Equation 13-1

PHOTOMETRY

At the same time, Bunsen contributed to the field of photometry by inventing an instrument known as the *Bunsen photometer*. The human eye is not able to make very accurate comparisons between two bright objects, but it can very accurately judge the relative brightness of two adjacent areas. In the Bunsen photometer, shown in Fig. 13-3, two lamps to be compared are set up at opposite ends of a slide. Between these lamps, a frame carries a piece of opaque paper made translucent in the center by a grease or oil spot. The disc is then slid back and forth until the grease spot seems to have the same brightness as the surrounding paper. When this balance is achieved, the distances between the paper and the two sources are measured. The relative strength of the two lamps can be calculated using the inverse square law relation as shown in Equation 13-1 of Fig. 13-3.

The Bunsen photometer was important because it was the first mechanism for quantitatively measuring the light output of a source. For a great many years, the standard of illumination was a candle.

This was defined by international agreement as a *spermaceti candle* (made from whale oil) burning at the rate of 120 grains per hour and viewed in the horizontal plane. Certain difficulties were involved in the direct use of the international candle. As the years went by, whale oil became more difficult to obtain. Keeping the candle burning at the precise rate was also difficult. To simplify measurements, various bureaus of standards calibrated incandescent lamps in terms of candlepower or equivalent candles. These were used as secondary standards. Nevertheless, the standard candle had a remarkable lifespan dating from the 1860s until 1940.

In 1940 scientists had planned to adopt an international standard light source. The adoption was held off until the end of

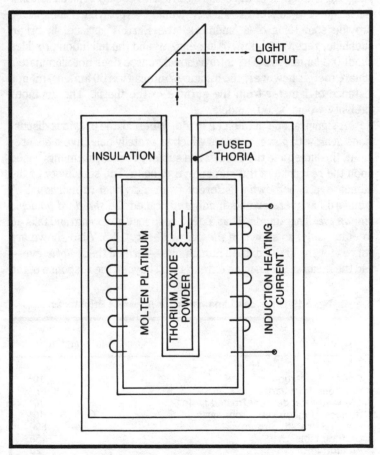

Fig. 13-4. The international standard light source.

World War II, however. This international source consists of a tube made of fused thorium oxide containing powdered thorium oxide. The tube is immersed in a pool of platinum which is maintained precisely at the freezing point of 1769° C. At this temperature, the thorium glows with a brilliant white light. This light was defined as being exactly 60 times the brightness of a standard candle. Induction heating is used to stir the platinum as well as to heat it, thereby assuring a uniform temperature in the vessel. See Fig. 13-4.

The modern standard of luminous intensity is usually expressed in *lumens per steradian* or *candela*. This sounds more complicated than it is. Imagine a sphere of one meter radius. The surface area of the sphere is 4π meters squared. If a 1-candlepower source were placed at the center of the sphere to radiate equally in all directions. Each square meter of the surface would be receiving 1 lumen. To provide something of a landmark, the Earth's surface, in bright sunshine, receives about 10^5 lumens/m^2 and the full moon provides about 0.3 lumen/m^2. The international source does not illuminate a square meter; however, the luminous intensity is 60 lumens/m^2 at a distance of 1 meter from the surface of the thoria. The luminous intensity would be 60 candela.

A significant point must be made. Whereas the previous discussions dealt with power in watts which is a totally objective measurement, the lumen and the candela are subjective measurements based upon the response of the human eye to light. The sensitivity of the human eye is somewhat different from individual to individual. A standard has therefore been adopted, called the *standard photopic human eye*. This standard has a peak response at 555 nm and falls off to either side, as shown in the curve of Fig. 13-5. Also shown are curves for the dark-adapted human eye—termed the *scotopic* eye— and the mean sun. Sunshine extends well beyond the response of the

Table 13-1. Various Landmarks for Apparent Brightness.

OBJECT	lm ster^{-1} m^{-2}
Crater of a carbon arc	10^9
Sun seen from earth	1.5×10^{10}
100 Watt incandescent frosted lamp	10^6
Surface of moon seen from earth	5×10^4
Hospital operating room	806
Drafting table	155
Auditorium	20

human eye in both the short and long wavelength directions. As shown from Equation 13-1 of Fig. 13-5, as the wavelength departs from 555 nm, it requires progressively more watts to give the same effective illumination. For example, 10 times as many watts per square meter at 480 and 640 nm are necessary to provide the same illumination as required at 555 nm. At 555 nm, 1 watt of radiative power is equivalent to 680 lumens.

To convert from candlepower to lumens, multiply by 4π. The candela is one lumen per steradian, or per $\frac{1}{4}\pi$ sphere.

Table 13-1 provides some landmarks for apparent brightness. For inspection purposes and medical applications of imaging fiberoptic devices, ratings between 300 and 900 are probably desirable.

THE RADIATION LAWS

In about 1880 an Austrian physicist, Joseph Stephan, experimentally deduced that the radiation from a heated body was pro-

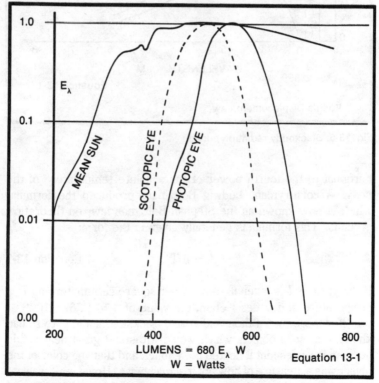

Fig. 13-5. Luminous responses.

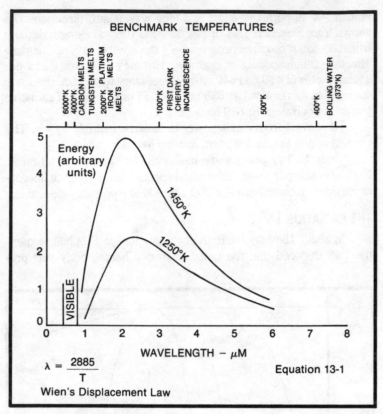

Fig. 13-6. Blackbody radiation curves.

portional to the fourth power of the absolute temperature of the body. A countryman, Ludwig Boltzman, produced the formula, which is now known as the Stephan-Boltzman law, on theoretical grounds. This formula is generally given in the form:

$$I = \delta T^4 \qquad \text{Equation 13-1}$$

If the value of I is given in calories per square centimeter and T is temperature in degrees Kelvin, the value of $\delta = 1.36 \times 10^{-12}$.

For more than a thousand years, blacksmiths and potters had known that solid objects which would withstand great heat would become incandescent if heated hot enough and that the color of the object was a measure of how hot it actually was. However it was not until the 1880s that extensive data became available on the spectra of heated objects and the measurements of the quantity of light emitted were made. The discovery was then made that the spectra of most solids such as iron and glass were essentially identical at any given temperature, but the shape of the curve changed with the peak shifting upward as the temperature increased. The curves of Fig. 13-6 show the curves of spectrum of two different temperatures, 1250 and 1450 degrees Kelvin. In 1895 the Wien's displacement law describing the shift of the peak of radiation was announced. This formula is shown in Equation 13-1 of Fig. 13-6.

Materials have to get up to very high temperatures before very much of their radiation is in the visible region. This was not fully appreciated when Edison commenced his series of experiments on the incandescent lamp in 1887. From the data at the top of the curve in Fig. 13-6, which was calculated using Wien's displacement law, you can see that the difficulty with the electric incandescent lamp is that the filament must be operated at temperatures where very few materials still have any remaining strength. The melting points of iron and platinum are far too low to give a good light in the visible region. Only tungsten and carbon have melting points high enough to present any significant efficiency in illumination. A modern tungsten lamp built for household use in the 100-watt size will deliver about 16 lumens per watt. The filament operates at something like 3100° K. The only other conductive metals which come close to offering this type of performance are rhenium and osmium. It is small wonder that earlier investigators failed to produce a satisfactory long-lived incandescent lamp. Only the persistence of Edison in trying more than 14,000 different lamp filaments made a suitable incandescent lamp filament possible.

Even such refractory materials as carbon and tungsten have the visible segment of the output curve on the *descending* slope of their spectrum. For photoflood and optical projection purposes, lamps are typically operated at higher temperatures than for ordinary illumination purposes. You can see from the curve that this will tend to greatly increase the light output and flatten the curve; however, this is accomplished only at the expense of greatly diminished filament life. A reduction of six percent in the supply voltage will double the projected life of a lamp or tube filament, and an increase of six percent will about halve the life. For applications like optical instruments, it is not unusual to operate a lamp at 24 percent above its rated voltage so that an initial 3200-hour life is shortened to 400 hours.

Increasing the input voltage to the lamp does not proportionally increase the input power. The power dissipated in a resistance is equal to E^2/R; however, the increased temperature of the filament will result in an increase in R. Therefore, raising the voltage by 24 percent will not increase the input power by 53 percent but rather something more like 33 percent. Because of the fourth power temperature law and the shift in peak frequency, the number of lumens can be doubled. From the standpoint of electrical efficiency, the high-voltage short-life operation is superior to the normal cooler operation. And it certainly does produce more light.

THE ATOM

At the time of Edison's invention of the electric light in 1879, the world of chemistry and physics still was bound to the atomic theory introduced by the English chemist, John Dalton, in 1803. The concept of the *atom* had originally been introduced in 400 B.C. by Democritus and the word was chosen for the meaning *noncuttable* or *indivisible*. Democritus had held on purely philosophical grounds that the universe must ultimately consist of identical particles that were themselves indivisible. The atoms of Democritus were all alike, and all material was held to consist of atoms in different combinations and arrangements. This theory had fallen into disuse by the time of Dalton with the discovery of chemical elements that were different from one another and could not be further divided by any chemical means. Dalton held that the atoms of each element were all identical but that they differed from element to element but that there was no further division possible, once the elemental stage had been reached.

By 1883 Johann J. Balmer, a Swiss physicist, had come upon a peculiar relationship concerning the lines of the Fraunhofer series. It had been found by experiment that the lines labeled C, F, G, and H were caused by the absorption of components in sunlight created by tenuous hydrogen in the solar atmosphere. Balmer discovered that there was a harmonious relationship in these lines that could be described by the relationship:

$$\lambda = 364.6 \frac{N^2}{N^2 - 4} \text{ nanometers} \qquad \text{Equation 13-2}$$

When N took the value of 3, the equation yielded a wavelength of 656.28, the wavelength of the C line. When N was 4, 5, or 6, the wavelengths were 486.13, 434.05, and 410.18 nanometers, which matched the F, G, and H lines.

The Balmer equation is only a descriptive one. It has nothing to do with explaining the cause of the spacing of the spectral lines. However, the match between the result and the physical facts seems to be too great to be a mere coincidence. The discovery of such a pattern in the spectral lines for hydrogen made it seem likely that similar patterns could be found elsewhere. It was indeed a major clue. There was a subtle order which could be discovered somewhere in the chaos of data.

THE GAS THERMOMETER

Prior to 1890, it had been discovered that there was more to the spectrum than meets the eye. In the attempt to quantify the amounts of radiation in various portions of the spectrum, a number of spectroscopists had used the *gas thermometer*. This device is shown in Fig. 13-7. A piece of glass tubing is blown with a bubble at the end. If the tubing is warmed slightly above room temperature, a drop of colored water can be placed on the end of the tube. As the tube cools, the drop will be sucked into the tube by the contraction of the air

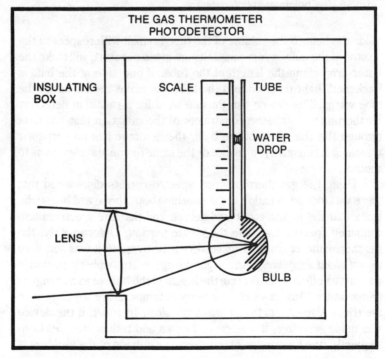

Fig. 13-7. The gas thermometer photodetector.

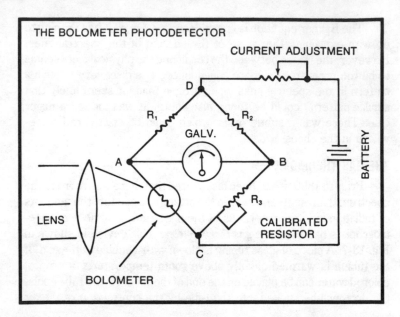

Fig. 13-8. The bolometer photodetector.

inside the bulb. If the volume of the tube is small with respect to the volume of the bulb, a very small change in temperature will make the water drop climb the length of the tube. If one side of the bulb is blackened, light directed into the tube will make the air inside the tube warm. The device can thus be used for a radiation detector. Furthermore, within the constraints of the radiation that will pass through the thin glass of the bulb, the detector has no particular wavelength sensitivity; it responds the same to one wavelength as to another.

Using the gas thermometer, spectroscopists discovered that there was indeed radiation in the sunshine both above and below the limits that the human eye could detect, and that the spectrum also contained spectral lines like the visible portion. Unfortunately, the gas thermometer could be made very sensitive, but it was sensitive to *just about everything.* Any slight change in atmospheric pressure was just as effective in altering the height of the bubble as a change in temperature. Any slight change in room temperature would also do the trick. Also, the detector was very slow. In short, if the device was made sensitive, it was cantankerous and tedious to use. Consequently, the knowledge of the spectrum outside of the visible was rather limited.

THE BOLOMETER

In 1890, Samuel Pierpont Langley produced the first really satisfactory photo detector with his invention of the *bolometer*. In this detector, shown in Fig. 13-8, a tiny resistor made of extremely fine platinum wire is used as one leg of a wheatstone bridge. Platinum has the advantage that it can be drawn into very fine wires which have a very stable and well established variation of resistance with temperature. Platinum changes resistance by 0.003 ohm/°C and has a resistivity of 10.5×10^{-6} ohm cm at 20°C. If all of the resistors of the wheatstone bridge are made of platinum and if they are all equal, there will be no potential difference between points A and B and the galvanometer will not deflect. Let us suppose, however, that incident radiation upon the bolometer resistor raises its temperature. In this case, the voltage drop from A to C will be larger than the voltage drop from B to C, and the galvanometer will deflect.

For example, suppose that at balance, all of the resistors in the bridge are 200 ohms and the voltage from D to C is 2 volts. Next suppose that the temperature of the bolometer alone is raised by 1 degree Celcius. The bolometer resistance will change by 0.003 × 200 × 1°C, or 0.6 ohms so that the bolometer resistance is 200.6 ohms. In this case, the voltage from A to C will be 1.003 volts, whereas the voltage from B to C will be 1.000 volts. The bridge has a sensitivity of 3 mV/°C. The calibrated resistor could be readjusted to renull the bridge, and the temperature rise could be calculated.

The bolometer has the advantage that the ambient temperature is canceled out by the variation in the other resistors of the bridge. It also has a distinct advantage compared to the gas thermometer in that it can be directly calibrated in power sensitivity. With no radiation input to the bolometer, the resistance of the bolometer with respect to power dissipated can be determined by varying the voltage from D to C and rebalancing the bridge. The current flowing through the bolometer will warm it slightly, and its resistance will rise. The power dissipated will then be calculated from the resistance and the input voltage. This direct calibration can then be used because the number of watts required to produce a given change in resistance has been determined and the optical power input can be read directly in watts.

With the advent of the bolometer, the interest in nonvisible spectroscopy increased and continuing research led to the identification of spectral lines outside the visible region for a number of materials. Hydrogen continued to be a prime topic for study because more was known about the spectrum. Professor Theodore Lyman contributed a formula for the lines in the extreme ultraviolet, and Friedrich Paschen did the same for the infrared portion of the spectrum. Interestingly, it was found that the Balmer, Lyman, and Paschen series could be made to harmonize with one another in a single equation. This revised equation is:

$$\nu = \frac{1}{\lambda} = 109{,}678 \left(\frac{1}{M^2} - \frac{1}{N^2} \right) \quad \text{Equation 13-3}$$

$$= \frac{1}{\lambda} \, cm^{-1} = \text{WAVE NUMBER}$$

The wave number is usually given as the reciprocal of the wavelength in centimeters. The equation works out the various series shown in Table 13-2.

Note that N must always be at least one greater than M or the quantity in the brackets would be zero or negative. Table 13-2 gives only the longest wavelength line in each series. There is a line for every value of N greater than M + 1. The bottom two series are much more recent than the top three and lie in the far infrared.

As the evidence accumulated for the order in the series of spectral lines, there was still little to indicate the reason for these lines to exist. Why should hydrogen gas under low pressure behave in such a complicated manner when most solid materials simply gave off a continuous spectrum?

THE EDISON EFFECT

A very substantial clue was to come first from the work of Edison and then from the work of Sir Joseph Thompson. In 1893, Edison had been working upon a lamp in which he had installed a small parabolic reflector to focus the light in one direction. Edison discovered that a small electric current would flow between the plate and the filament when the plate was positive with respect to the filament but not when it was negative. Edison had invented the thermionic diode! However, Edison was always a pragmatic businessman and he really did not see any great use for this phenomenon. It was patented, announced, and promptly dropped, but fortunately not by everyone. The *edison effect* was investigated by several other people.

From the early 1800s it had been known that when most of the air was pumped out of a glass tube, that tube could be made to light up if a high voltage were applied between electrodes sealed in the

Table 13-2. Equation 13-3 Results of Various Series.

SERIES	WAVELENGTH (first line of series)	M	N
Lyman	121.57×10^{-9}m	1	2
			3
			4 ,,,
Balmer	656.47	2	3
			4
			5 ...
Paschen	1,875.6	3	4
			5
			6 ...
Brackett	4,052.3	4	5
			6
			7 ...
Pfund	7,459.9	5	6
			7
			8 ...

tube. In fact, most of the hydrogen spectra had been obtained in this manner. In 1878, Sir William Crooks discovered that if the electrodes were made with particular shapes, they would cast shadows on the walls of the tubes. The fact that the current carrying items would miss the target indicated they were particles with some mass. Furthermore, Crooks was able to show that the particles could be deflected by a magnet.

X-RAYS

The next steps in the drama followed swiftly upon the Edison invention. In 1895, Wilhelm K. Roentgen discovered X-rays and an apparatus to make them. The X-ray came into almost instant use since it could be made using an electrostatic generator and a Geissler tube, both of which could be found in practically any self-respecting physics lab.

The other step came from Joseph J. Thompson. In a series of experiments conducted between 1895 and 1897, Thompson investigated the properties of the current-carrying particles that were later to be named *electrons*. Thompson was able to prove that these had mass and that the energy of the electrons was a function of the potential difference across the electrodes. Furthermore, Thompson studied the rate at which magnetic and electric fields bent the trajectory of the particles. From this, he was able to determine the ratio of their mass to their charge. He was also able to show that the charge was negative. The evidence was beginning to mount that atoms were not indivisible after all. It seemed as though one could pull these electrons out of nearly anything, and they always were the same, no matter where they came from.

It had been discovered in 1888 that a freshly polished zinc plate would lose an electrical charge when it was illuminated with ultraviolet light, but only if the charge was negative. As the study progressed, other metals, particularly the *alkalai metals* such as lithium, sodium, potassium, rubidium, and caesium all possessed this property of losing charge upon illumination. Furthermore, it was discovered that some of these metals, particularly caesium, would lose charge with light in the visible range.

By 1898 it had been established that the metals were losing the charge because they were emitting electrons. The most peculiar part about the phenomenon had to do with the details of the emission. First, it was found that for each of the metals there was some maximum wavelength of light that would cause the effect. Secondly, for any given single wavelength which could cause the effect, all of the emitted electrons had the same energy. Increasing the amount of light only increased the number of electrons, but the energy of the electrons was inextricably bound to the wavelength or frequency of the light; the higher the frequency, the higher the energy. It further seemed that for every metal, there was some minimum energy where the electrons were no longer emitted.

This was the first serious flaw with the wave theory of light. The photoelectrons seemed to be telling us that somehow, contrary to the wave theory, there was some energy function uniquely associated with the frequency of the light. Furthermore, this energy was measurable and seemed to be linearly related with frequency. In the wave theory of light, there was nothing to connect the wave with some minimum amount of energy.

Einstein pointed out in a beautifully clear example that sand is normally considered a continuous medium. You can purchase sand in any amount—a 100-pound bag, a cubic yard, etc. And yet, in the final analysis, sand is made of grains. You cannot purchase any less than a single grain of sand. If sand were as precious as gold, we would probably be concerned with keeping track of each grain. Thus, it seems likely that other things such as light which seem continuous when used in large quantity might still be granular when considered on a sufficiently fine scale.

MAX PLANCK

It was just this sort of consideration which led Max Planck to publish his *quantum theory* of light in 1900. Planck noted that the photoelectron experiments indicated that the energy present in a *quantum* of light is given by

$$E = hf \qquad \text{Equation 13-4}$$

If E is energy in ergs, f is frequency in hertz, and h is a constant with the value of 6.63×10^{-27} erg-sec. The symbol ν is frequently used to represent the frequency; however, this is usually given as wave number or reciprocal of wavelength, whereas frequency in hertz is 3×10^8 m/sec/λ in meters. This requires an adjustment in the constant h.

From this, it should be evident that quanta are very small grains of energy indeed and that they are smaller for lower frequencies than for high. For example, the wavelength for the peak response of the human eye of 0.555 μm corresponds to a frequency of 5.41×10^{14} Hz compared to 10^6 Hz for the center of the broadcast band. The quanta for visible light are some 10^8 times as energetic as those for broadcast radio.

The Planck equation for the first time provided an explanation of the shape of the radiation spectra as shown in Fig. 13-6. Planck considered that the heated solids consist of an ensemble of tiny oscillators, each with its own frequency. He provided a mechanism for calculating the amount of energy to be released by a heated black body over any interval of wavelength.

In 1905 Einstein published his *photoelectric equation,* which states that the maximum energy of the photoelectrons expelled from the alkalai metals is given by:

$$\frac{1}{2} mv_m^2 = hf - w \qquad \text{Equation 13-5}$$

where m = mass of electron, v_m = maximum velocity of electron, and w = energy required to break the electron loose.

For the above equation, h and f are defined as in Equation 13-4.

THE FLEMING VALVE

In 1904 the British scientist, John A. Fleming, introduced the *Fleming valve* or thermionic diode for use as a detector in radio work.

This was really simply an Edison effect diode with certain refinements to make it suitable for radio detection. In 1907, the American scientist, Lee H. De Forest, introduced the control grid into the thermionic diode and produced the thermionic triode tube which was capable of amplifying signals. This event marks the beginning of electronics.

In 1907, the American physicist, Robert A. Milliken, performed an experiment in which very fine drops of oil were levitated by an electric field. Milliken had developed a technique for weighing the oil drops by noting the rate of drift and found that the fields required to levitate the drops varied in small but discrete steps. He deduced that the steps were due to differences in the charge on the drops which corresponded to single electrons. He was thus able to calculate the charge of the single electron. This charge is also very small and is equal to 1.60×10^{-19} coulombs. A $1\text{-}\mu F$ capacitor charged to a potential of 1 volt contains a charge of 6.25×10^{12} electrons!

In gaseous discharges, it was found that the electrons were accelerated in one direction and the positive charges were accelerated in the other. With the same deflection techniques used for the electron, the mass of the electron was calculated as only 1/1840 the mass of the hydrogen atom. Also, the charge of the positive portion was the same but of opposite sign as that of the electron. In 1911 Ernest Rutherford arranged an experiment to bombard some extremely thin gold targets with a beam of positively charged particles that were obtained by an electric discharge in very rarified hydrogen. Rutherford found that most of the protons or positive projectiles passed straight through the gold sheet, but a few rebounded as though they had struck something very solid. He was able to deduce from this that most of the space inside the atom was empty and that most of the mass was concentrated in the very center of the atom. He was also able to conclude from the nature of the rebound that the center of the atom was positively charged.

The idea of the structure of the atom similar to the structure of the solar system had achieved some considerable coinage. After all, there were the opposite charges of the electron and the nucleus; without some action to hold them apart, the oppositely signed charges would pull together. What better way than centrifugal force, which held the earth in a stable orbit from the sun? Unfortunately, this theory had a major flaw. By the turn of the century, the theory of electromagnetism and electromagnetic radiation was sufficiently well developed to state with fair certainty that the electron would be radiating energy at the frequency of its orbit about the nucleus. This energy loss would cause the orbit of the electron to decay and the electron would ultimately crash into the nucleus. A second difficulty lay in the fact that the best understood atom, hydrogen, would yield radiation only in discrete spectral lines and not in a continuum when it was deprived of an electron by an electrical discharge. This seemed to imply that there were certain preferred orbits for the electrons in the hydrogen atom and other atoms as well. In the simple mechanical view of a very light electron orbiting a proton, there were no preferred orbits. Each orbit simply required a slightly different energy of the electron, but it presented a continuous scale.

NIELS BOHR

In 1913 Niels Bohr announced a new theory in which the atom was constructed of a nucleus and a number of electrons arranged around the nucleus in shells of increasing radii. There was but a single electron in the case of hydrogen. However, there were a number of available shells. The shells each represented a preferred state of the electron. If the electron were removed from the lowest shell by something like an electric discharge, it could fall from shell to shell, each time giving up a photon of light corresponding to the energy difference between the shells. Within the shell itself, however, the electron would lose no further energy. Eventually the electron would tend to wind up in the shell having the lowest available energy. The various spectral lines to be seen from an electrical discharge in hydrogen represented the energy difference between various shells. The fact that some lines were stronger than others means that the probability of a transition between two particular shells with that energy difference is less.

This description of the Bohr atom is of necessity oversimplified, but it is adequate to provide a feeling for some of the discussions on photo sources and photodetectors to follow. In essence, the theory accounted for the radiation energy loss by ignoring it. On purely pragmatic grounds, atoms do not eventually decay from radiation even though it can be proven in a macroscopic lab experiment that an electron whirled in a circle *will* create a magnetic field along the axis of rotation and *will* radiate waves at the frequency of rotation which *will* be polarized in the plane of the rotation. This form of rotationally induced radiation is commonly detectable as synchrotron radiation in the laboratory from circular electron accelerators and in the radiation from certain type stars. The chief virtue of the Bohr atom was that it provides a mathematical model for predicting the spectra to be obtained from various discharges in different materials.

In the intervening years, our knowledge of atomic structure has considerably altered. The Bohr model is used today principally as a simplified teaching tool to understanding the phenomena involved. In 1927, the American physicist, Clinton J. Davisson, and the British physicist, George Thompson, independently discovered that electrons can be made to demonstrate wave properties such as single-hole and two-slit diffraction. This leaves us with the peculiar situation that both matter and light can sometimes be shown to have properties associated only with waves and sometimes can be shown to have properties associated only with particles.

14

Photodetectors

In Chapter 13 two different types of photodetectors, the gas thermometer and the bolometer, were examined. Both of these detector types operate by using the incident radiation to heat something and then measuring the heating effect. The gas thermometer can be made very sensitive but is subject to all sorts of errors from outside sources. On the other hand, the bolometer is much less subject to outside influences, other than the desired radiation measurement; however, it has a tendency to be slow, although not as painfully slow as the gas thermometer but usually limited to responses less than a few kilohertz. This is far too slow for most communications applications. A detector which does not respond by warming up and then cooling off is required for practical communication rates. The use of the bolometer for radiation measurements is generally confined to power metering and calibration of other, faster detectors for which a solid relationship between the incoming radiation flux and the output response is not known.

BACK TO THE ATOM

In the Bohr theory of the atom, the electron was restricted to certain energy levels and was required to move in orbits at fixed radii about the nucleus. These levels could be quantified and identified by a quantum number. The Bohr model of the atom is shown more or less schematically in Fig. 14-1. In falling from orbit 4 to orbit 3, the electron would give up a photon of energy determined by the Equation 14-1 of the figure.

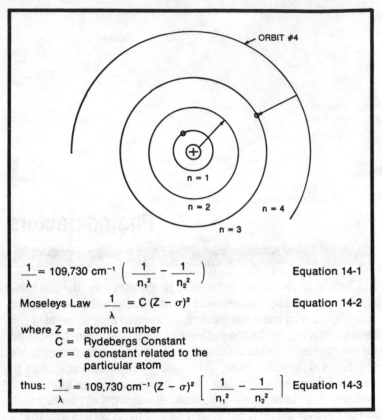

ORBIT #4

$n = 1$
$n = 2$
$n = 4$
$n = 3$

$$\frac{1}{\lambda} = 109{,}730 \text{ cm}^{-1} \left(\frac{1}{n_1^2} - \frac{1}{n_2^2} \right)$$ Equation 14-1

Moseleys Law $\dfrac{1}{\lambda} = C(Z - \sigma)^2$ Equation 14-2

where Z = atomic number
C = Rydebergs Constant
σ = a constant related to the
particular atom

thus: $\dfrac{1}{\lambda} = 109{,}730 \text{ cm}^{-1} (Z - \sigma)^2 \left[\dfrac{1}{n_1^2} - \dfrac{1}{n_2^2} \right]$ Equation 14-3

Fig. 14-1. The Bohr model.

The Bohr equation agrees very well with experimental results
for hydrogen. Unfortunately, it does not work out with atoms of a
more complex structure. In subsequent work done with the X-ray
spectra, Henry G. Mosely found that the wavelength for a particular
line depended upon the atomic number of an element and *not* upon its
atomic weight. The relationship between wavelength and atomic
number is known as Mosely's Law and is given by the expression of
Equation 14-2 of Fig. 14-1. When this is applied to the Bohr equation,
the result shown in Equation 14-3 of Fig. 14-1. For hydrogen, $Z=1$
and $\sigma=0$. This reduces to Equation 14-3 of the figure, and will
calculate the line wavelengths in the infrared, visible, and ultraviolet
regions. For molybdenum, $Z=42$ and $\sigma=0.5$ will suffice for calculating the line in the $n=2$ to $n=1$ transition as 0.071 nm.

For atoms with more than one occupied shell, the optical spectrum generally is derived from an electron falling into the outermost shell. It is this outermost shell which also determines the chemical properties of the atom. In contrast, the X-ray spectrum is usually caused by an electron falling into the innermost shells. Studied together with the chemical properties, the X-ray spectra permit a determination of the number of electrons which can occupy any given shell. The first ($n=1$) shell can have 2 and the second shell can have 8, etc.

The more modern *wave mechanics* treatment of the atom yields a series of wave functions which are usually designated by the Greek letter psi (Ψ), each of which is characterized by a set of three integral quantum numbers. Each wave function describes a certain region about the nucleus in which the electron can move. Each region has a certain energy level and is referred to as an *orbital* to distinguish it from Bohr's orbits. According to wave mechanics, the energy levels in the atom are composed of one or more orbitals. Atoms which have more than one electron have a distribution of electrons determined by the number and kind of energy levels which are occupied.

An in-depth study of the makeup of these wave equations is not within the scope of this text; however, it is necessary that some insight into the assignment of the terminology be given because the literature describing the devices to be used in fiberoptic engineering frequently makes reference to some of the terms from the wave mechanics view of the atom.

THE QUANTUM NUMBERS

☐ The Principal Quantum Number—n

The energy levels in the atom are arranged into shells as determined by the principal quantum number n. This is similar to the Bohr theory, and the values of n are integers starting at 1 and proceeding to infinity. The shells are sometimes lettered starting with the letter K, thus: $K=1$, $L=2$, $M=3$ . . . . .

☐ The Azimuthal Quantum Number—l

The letter l is frequently used to designate this quantity which stems from the wave mechanics prediction that each main shell is composed of one or more subshells. The subshell numbers may have any integral value up to one less than the principal quantum number "n". Thus for $n=1$, $l=0$, and $n=2$, the value of l can be 0 or 1, etc. The number of subshells in any given shell is simply equal to n; however, the first subshell is labeled zero.

In the spectra of the alkalai metals lithium through

caesium, a group of spectral lines were originally labeled *sharp, principal, diffuse,* and *fundamental.* These names have lent themselves to the first four azimuthal quantum numbers. Thus $s=0$, $p=1$, $d=2$, and $f=3$. For values of l greater than 4, the listing simply continues up the alphabet.

The first four subshells are of principal interest in most cases since these represent the lowest energy or *ground state* of the atoms. The higher orders describe *excited states* or states which the atom will normally attain only when subjected to outside excitation due to electric fields, extreme or ionizing heating, etc.

To describe a subshell within a given shell it is normal to write the value of n followed by the designation of the subshell. For example the s subshell of the second shell ($n=2$; $l=0$) would be written 2s. For $n=$ and $l=1$, this would be the 2p subshell. This is one of the designations frequently used in solid-state device literature.

□ The Magnetic Quantum Number—m

Each subshell is made up of one or more orbitals that are designated by the letter m. The magnetic term is used to explain that additional lines appear in the spectrum when the atom emits light within a strong magnetic field. This property is sometimes referred to as *Zeeman splitting* of the spectral lines. The value of m can be equal to any value between $+l$ and $-l$, including zero. For instance, for $l=3$ there are seven (-3, -2, -1, 0, 1, 2, and 3). Note that all of the s subshells have only a single orbital because the value of $l=0$.

Figure 14-2 shows the energy distribution obtained from the wave equations for various lower order orbitals. The energy increases with an increasing principal quantum number, and the spacing in energy between the higher quantum numbers is smaller than the spacing between the lower numbers. Furthermore, beyond the $n=3$ level, there is a possibility for an overlap in the energy of the levels. In particular, 3d has a higher energy than 4s. This overlap possibility increases with higher values of n.

In Fig. 14-2, the various values of m have been shown as dashes. The s subshell always has only one value of m, as previously noted. Each of these dashes represents one orbital.

There is yet another number used in the system. This is the *spin number.* The electron behaves as if it were spinning about some axis with the charge describing a small circle. The moving charge will generate a magnetic field. This spin has the convention that the direction of the magnetic field can be only up or down, and the spin is assigned a quantity of either $-\frac{1}{2}$ or $+\frac{1}{2}$. The *Pauli exclusion*

Table 14-1. Number of Orbitals and Electrons Versus Subshells.

Subshell	Number of orbitals	Number of electrons
s	1	2
p	2	6
d	5	10
f	7	14

principle states that no two electrons in a given atom may have all four quantum numbers the same. For example, select the 1s orbital. For this orbital, $n=1$, $l=0$, and $m=0$. There are still two electrons, but they must have different spin. This limits the number of electrons in any orbital to two, since there are only two possible values of spin. The Pauli exclusion principle also limits the number of electrons that can be accommodated in the s, p, d, and f subshells as shown in Table 14-1. The number of electrons in any shell is limited to $2n^2$. Thus for $n=2$, the maximum number of electrons would be eight. Of

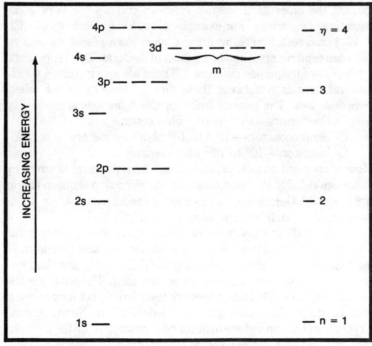

Fig. 14-2. Energy diagram for the lower orbitals.

241

these, two would be found in the s subshell and six in the p subshell.

On some energy diagrams, it is common to designate the spins of the electrons with arrows over the em-dashes. In this notation, an up arrow designates a $+\frac{1}{2}$ spin, and a down arrow designates a $-\frac{1}{2}$ spin number.

CONDUCTION IN METALLIC CRYSTALS

Most of the commonly used conductors of electricity are metallic solids in crystalline form. A few, such as mercury, are liquid at room temperature. However, more of the common ones are solid and crystalline at room temperature. These crystals generally behave as if the positive ions (which consist of the nucleus of the atoms and the inner orbitals) are locked into the lattice of the crystal and the outer electrons belong to the crystal as a whole rather than to any single atom. This sort of "sea of electrons" is relatively free to move. If an electron is withdrawn from one end of the crystal and a replacement is inserted at opposite ends of the crystal by an external electrical circuit, this sea quickly comes back to equilibrium. For this reason, metallic crystals are generally good conductors of electricity.

On the other hand, certain types of crystals are very good insulators of electricity. For example, silver has a resistivity of 1.62×10^{-6} ohm centimeters, whereas carbon has a resistivity of 5×10^{14} ohm centimeters when in the form of diamond. This represents a 20 order-of-magnitude difference. There are also materials which have resistivities in between these limits. These items are called *semiconductors*. The general limits for the definitions are:

☐ Conductors—10^{-6} to 10^{-4} ohm centimeters
☐ Semiconductors—10^{-4} to 10^9 ohm centimeters
☐ Insulators—10^9 to 10^{25} ohm centimeters

Most of the solid metallic crystals range from the level of silver to plutonium at 1.5×10^{-4} ohm centimeters. Silicon is a semiconductor at 8.5×10^{-2}. Germanium is much more conductive at 4.5×10^{-5}. Both are in crystals at room temperature.

As far as the chemistry of most atoms is concerned, it is only the outermost orbitals which become involved in chemical reactions. If the outermost orbitals are not completely filled, the atom can exchange electrons from the outer or *valence shell*. The noble gasses have a completely filled outer shell and therefore do not tend to react with anything. The noble gasses include Helium, Neon, Argon, Krypton, and Xenon and are to be found in column "0" on the periodic table of Table 14-2. If a descending staircase is drawn on the periodic

Table 14-2. The Periodic Table of Elements.

PERIOD	1A	2A	3B	4B	5B	6B	7B	8	8	8	1B	2B	3A	4A	5A	6A	7A	0
1	+1 −1 H 1																	0 He 2
2	+1 Li 3	+2 Be 4											+3 B 5	+2 +4 C 6	+1 +3 +5 −3 N 7	−2 O 8	−1 F 9	0 Ne 10
3	+1 Na 11	+2 Mg 12											+3 Al 13	+2 +4 Si 14	+3 +5 −3 P 15	+4 +6 −2 S 16	+1 +5 +7 −1 Cl 17	0 Ar 18
4	+1 K 19	+2 Ca 20	+3 Sc 21	+2 +3 +4 Ti 22	+2 +3 +4 +5 V 23	+2 +3 +6 Cr 24	+2 +3 +4 +7 Mn 25	+2 +3 Fe 26	+2 +3 Co 27	+2 +3 Ni 28	+1 +2 Cu 29	+2 Zn 30	+3 Ga 31	+2 +4 Ge 32	+3 +5 As 33	+4 +6 −2 Se 34	+1 +5 −1 Br 35	0 Kr 36
5	+1 Rb 37	+2 Sr 38	+3 Y 39	+4 Zr 40	+3 +5 Nb 41	+6 Mo 42	+4 +6 +7 Tc 43	Ru 44	Rh 45	+2 +4 Pd 46	+1 Ag 47	+2 Cd 48	+3 In 49	+2 +4 Sn 50	+3 +5 Sb 51	+4 +6 −2 Te 52	+1 +5 −1 I 53	0 Xe 54
6	+1 Cs 55	+2 Ba 56	♦ 57–71	+4 Hf 72	+5 Ta 73	+6 W 74	+4 +6 +7 Re 75	+3 +4 Os 76	+3 +4 Ir 77	+2 +4 Pt 78	+1 +3 Au 79	+1 +2 Hg 80	+1 +3 Tl 81	+2 +4 Pb 82	+3 +5 Bi 83	+2 +4 Po 84	−1 At 85	0 Rn 86
7	+1 Fr 87	+2 Ra 88	★ 89–103															

GROUP

LIGHT METALS

HEAVY METALS

BRITTLE

DUCTILE

LOW-MELTING

NON METALS

INERT GAS

TRANSITION ELEMENTS (BETWEEN GROUPS 2A AND 3A.

	3B													
♦ LANTHANIDES (RARE EARTHS)	+3 La 57	+3 +4 Ce 58	+3 Pr 59	+3 Nd 60	+3 Pm 61	+3 Sm 62	+3 Eu 63	+3 Gd 64	+3 Tb 65	+3 Dy 66	+3 Ho 67	+3 Er 68	+3 Tm 69	+3 Yb 70	+3 Lu 71
★ ACTINIDES	+3 Ac 89	+4 Th 90	+5 Pa 91	+4 +6 U 92	+4 +6 Np 93	+4 +6 Pu 94	+4 +6 Am 95	+3 Cm 96	+3 Bk 97	+3 Cf 98	Es 99	Fm 100	Md 101	No 102	Lw 103

OXIDATION NUMBERS

```
{ +2
{ +4
```

Si
14

+4 ── OXIDATION NUMBER
SYMBOL
ATOMIC NUMBER

KEY TO CHART

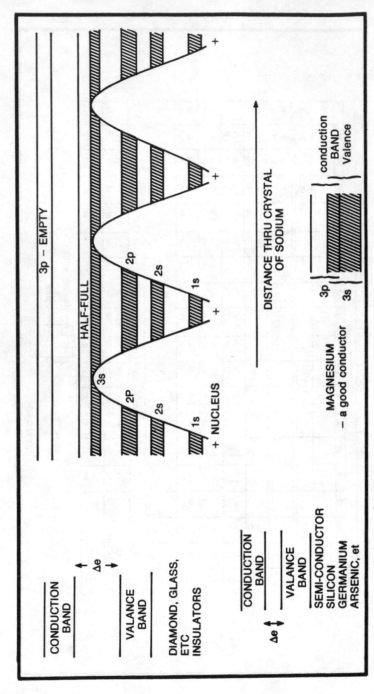

Fig. 14-3. Energy band picture of sodium.

table starting between boron and aluminum and descending between aluminum and silicon, horizontal between silicon and germanium, descending again between germanium and arsenic, you would find that the metals lie to the left of the staircase and the nonmetals to the right. Some of the elements in the immediate vicinity of the staircase are the semiconductors.

In order to explain the large—indeed, very, very large—differences in the conductivity of atomic crystals, the *band theory* of solids was evolved. This theory explains that in a solid, an energy band is composed of a large number of closely spaced energy levels. These energy levels are formed by the combining of atomic orbitals of similar energy within the crystal. For example, the ls orbitals in sodium combine to form a single 1s band that extends throughout the solid. The same is true of the 2s and 2p orbitals. However, the 3s orbital in sodium is only half-filled and the 3p and higher orbitals are empty.

One of the ways of illustrating the situation is shown in Fig. 14-3. In the figure, the nuclei are shown as the circled plus charges and the energy bands that are filled are shaded. Electrons, cannot freely swap or flow between the filled bands which are down in a pit between the atoms. However, the half-filled nature of band 3s extends through the crystal, and the electrons can happily swap from atom to atom in this band. Electricity flows through the crystal because of the unfilled band and because sodium crystals have very low resistivity.

The band containing the outermost electrons in the atom is called the *valence band* since this is the band which participates in chemical reactions. Any unfilled band which extends throughout the crystal is termed the *conduction band* since this is the band which gives the crystal its electrical conductivity. In sodium, the conduction and valence bands are the same. In magnesium, on the other hand, the 3s band is filled and cannot be used to transport electrons. However, magnesium has the property that the 3p band actually overlaps the 3s band. This means that electrons can hop between the 3s and the 3p bands without acquiring much in the way of energy. Magnesium is therefore also a good conductor.

In an insulator, a large energy gap is between the conduction and the valence bands. Therefore, a very large amount of energy must be supplied to get each electron to jump between the valence and conduction bands.

In a semiconductor, the gap between the valence and the conduction bands is very small. Therefore, relatively small amounts of energy will cause the electrons to leap from band to band. The crystal conducts electricity, although it has a relatively high resistance.

These illustrations do not present anything real which could actually be seen by examining the crystal, not even with a sufficiently high-powered microscope. These are merely diagrams that symbolize what is going on in a crystal in terms of the wave mechanics theory. However, the energy bandgap labeled Δe in the diagrams is a very real item that can be measured. The purpose is simply to provide you with something to visualize for a process which, like an attitude or an odor, is not actually visible.

One of the characteristics of semiconductors is the fact that the closely spaced bands permit thermally excited electrons to make the jump with less energy. This gives rise to the fact that silicon and germanium decrease in resistivity with increasing temperature.

Thus far, the crystals have been theoretically perfect; that is, the lattice of the crystal is completely regular. There are several types of defects which can exist in a crystal. The *Frankel defect* occurs when the crystal has the correct amount of cations and anions, but one of the cations is out of place in the lattice, between the anions in the interstices. The second type of defect is the *Schottky defect* in which a cation or an anion may be completely missing from the lattice.

One of the most important types of defects arise when impurity atoms are introduced into the crystal in order to enhance the semiconductor properties of that crystal. In this case, a few of every million or so atoms of the crystal are replaced by atoms of a foreign substance. This provides changes in the behavior of the crystal.

THE P-TYPE SEMICONDUCTOR

A good example of this is to be found when germanium (Ge) is *doped* with a few parts per million of gallium (Ga). The Ge atom has in the fourth orbital the structure $4s^2 4p^2$, meaning that there are two electrons in the s and p shells. Its neighbor on the periodic table, gallium, has a close approximation of the same structure, except that in the fourth orbital, Ga has $4s^2 4p^1$. In other words, it is shy one electron. When a few Ga atoms are introduced into an otherwise pure Ge crystal, the crystal winds up with holes where the structure is shy an electron. Under the influence of an applied electric field, an electron from an adjacent atom can jump over to fill this gap, thereby leaving a hole at its point of origin. By this mechanism, the hole may propagate through the crystal and the conduction of the crystal is enhanced. Because the carrier is positively charged, this type material is described as a *p-type semiconductor*.

In the case of silicon, which has a third orbital structure of $3s^23p^2$, the doping can be done with aluminum ($3s^23p^1$) or boron ($2s^22p^1$) although the boron doping seems to be more easily accomplished. The silicon lattice with the electron vacancy is also a p-type semiconductor.

THE N-TYPE SEMICONDUCTOR

Another type of doping is important in semiconductors. Arsenic (As) with the fourth shell containing $4s^24p^3$ has one more electron than germanium. If a few atoms per million in a Ge crystal were replaced with As the crystal would have an excess of electrons. In this situation, an electron inserted into one end of the crystal by an external circuit will propagate by displacing the extra electron. Again, the crystal is rendered conductive. In this case, because the carrier is an excess electron and is basically negative in nature, the crystal is referred to as an *n-type semiconductor*.

For silicon (Si), the addition of phosphorous (which has a third shell containing $3s^23p^3$ compared to $3s^23p^2$ for Si) will supply the excess electron. However, silicon doped with arsenic seems to be the more common choice.

OTHER SEMICONDUCTORS

A series of other crystals can be used to make semicondoctors. For example, from the previous data, you might expect that a crystal composed of gallium and arsenic, depending upon the proportions, might prove to be either a p-type semiconductor or an n-type semiconductor. As a matter of fact, this is true. GaAs devices form an important class of semiconductors known for their very high electron mobility and high temperature capability. Indium and antimony also make good semiconductors because of the high electron mobility. In a way, this is to be expected from an inspection of the table of the elements shown in Table 14-2. All of the known type p doping elements lie in column 3A, and all of the known type n doping elements lie in column 5A. It is not too surprising that InSb should make a good semiconductor, just as GaAs does. Of course tin (Sb) is a good conductor and cannot be used to make a semiconductor device, nor for that matter can lead (Pb).

Crystals of chemical compounds are also common, such as table salt (NaCl) and sugar ($C_{12}H_{22}O_{11}$). However, a practical compounded semiconductor has not yet been produced.

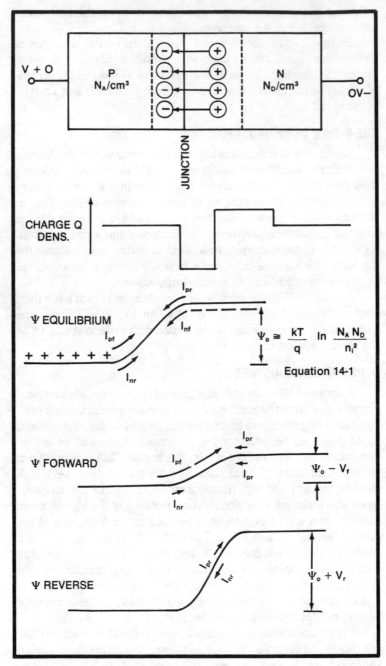

Fig. 14-4. The pn junction.

248

THE PN JUNCTION AND THE PHOTOVOLTAIC CELL

When a p-type semiconductor and an n-type semiconductor are placed in intimate contact, an interesting phenomenon takes place. Figure 14-4 shows this effect.

Assume that you have a pn junction in which the p portion has been more heavily doped than the n portion. The p-type carriers are holes and thus are acceptors of electrons. The number of such acceptors is usually designated by the symbol N_a. Conversely, the n portion donates electrons and the charge carriers are termed donors and designated as N_d. The heavier doping statement means that N_a is greater than N_d.

At equilibrium—that is, with no external circuit or no external applied voltage—the electrostatic imbalance in the vicinity of the junction proper will cause the excess electrons in an extremely thin layer adjacent to the junction proper to shoot the gap and fill some of the acceptors on the p side. In this condition, the electrons have piled up on the p side of the junction and some holes have developed on the n side of the junction.

In the next illustration down, the equilibrium potential, which is usually designated by the Greek psi major (Ψ), is shown. The area in which the carrier has been swept up is called the *depletion region* for relatively obvious reasons. In the equilibrium condition, a few carriers will be promoted in thermal energy enough to make the hill, but the number of forward and reverse carriers of either n or p will soon settle down to be equal and the charge will settle to a constant value. The voltage across this junction will be given by Equation 14-1 of Fig. 14-4 where k = Boltzman's constant − 8.61×10^{-5} eV/°K, T = Temperature in °Kelvin, q = Charge of an electron − 1.60×10^{-19} coulomb, and n_i = the number of carriers in the intrinsic material.

At a temperature of 300° K (27° C) or about room temperature, the value of kT/q is 0.026 volts. In much of this work, it is convenient to express energy in terms of electron volts or eV, which is the amount of work required to force an electron through a potential drop of 1 volt.

When a forward bias is applied to the junction from the external circuit, the potential barrier or hill is lowered and a greater number of the holes from the p-doped section can make the hill. In this case, the terminal voltage across the diode becomes equal to the equilibrium voltage, minus the forward voltage of the diode. The forward voltage is determined by the mobilities of the p and n carriers in their respective sections. In general, this is a linear function; that is, the diode acts to an external circuit very much like a resistor and a bettery in series. Once the external circuit has overcome the battery voltage, the change in current with a change in voltage is constant.

In the reverse biased mode—that is, when the p section is connected to a negative voltage supply and the n section is connected to a positive supply—the action of the external voltage is simply to increase the potential hill in the depletion region. The depletion region actually physically widens in this condition, and the diode takes on the characteristics of a capacitor. In practical diodes, there is always some small reverse current which does not depend strongly upon the barrier height. This reverse current in most diodes will double every 8° or 10° C. However, in a good quality diode at room temperature, the reverse current is very small. For all intents and purposes, there is no current flow in the reverse direction.

If the reverse bias is carried to a high enough level, the charge carriers can acquire enough energy to make hole-electron pairs by impact ionization. These new carriers can in turn generate more carriers, and the unit can break down in the *avalanche mode*. In this mode, the effective resistance of the diode will suddenly fall to a very low level. If the breakdown occurs at less than 6 volts in silicon or 3 volts in germanium, the breakdown is generally a *zener* breakdown due to tunneling or field emission rather than an avalanche.

Figure 14-5 shows a curve which depicts the typical performance of a good pn junction diode. In a test run on a series of 1N914 diodes from the parts drawer, it was found that the forward voltages (V_f) at 5 mA ran between 0.514 and 0.595 at 22°C and that the temperature coefficients ran between 1.58 to 2.13 mV/°C. The resistances in the forward mode were on the order of 20 to 30 ohms.

In the back biased mode, the diodes had to have much larger voltages applied. Before breakdown, the diodes showed a resistance in excess of 20 Megohms. However, at a voltage ranging from 105V for the worst and 150V for the best, the curve would sharply knee over and the effective resistance of the diodes would fall to a very low value—probably on the order of a few hundred ohms or less. In this case, the current had to be limited with a resistor and the voltage could be applied only in pulses to prevent destruction of the devices. The 1N914 is a typical silicon small-signal high-speed switching diode commonly used in digital work. For larger rectifier diodes, the forward resistance would be smaller.

THE PHOTODIODE: THE SOLAR CELL

One of the typical constructions for a solar cell employs a silicon crystal doped with arsenic to make it an n-type semiconductor. Onto this wafer there is deposited an extremely thin layer of p-type semiconductor, usually silicon doped with boron. Because the layer is so thin, its resistance is relatively high. It is therefore necessary to mask a contact pattern like a fern or other grid shape onto the p layer. When a photon of sufficient energy penetrates the p layer and gets to the junction, it can create a hole-electron pair which upsets the equilibrium value and causes a current to flow in the external circuit. Typically these diodes will develop something like 0.5V in bright sunshine and will deliver 0.6A per square inch (930 A/m²). The solar cell is relatively efficient in terms of electrical output versus light input but is very slow compared to a signal photodiode due to the large areas and high capacitances involved. In fiberoptic work, the use of solar cell type devices or photovoltaic cells is restricted to providing power for remote repeaters.

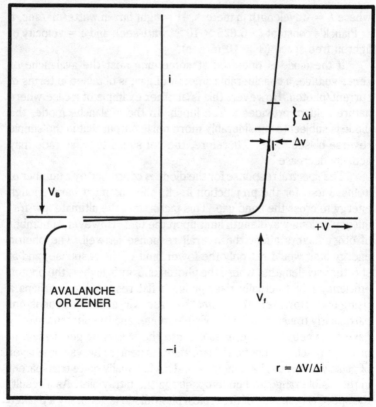

Fig. 14-5. The pn junction V/I curve.

251

THE REVERSE BIASED PN JUNCTION

The signal photodiode is usually a pn junction diode run in the reverse bias mode. In this case the diode—when maintained in the dark—will draw only the leakage or *dark current*. This dark current is relatively independent of the applied voltage below breakdown. When a photon arrives, if it has sufficient energy, it creates a hole-electron pair which must be swept up by the reverse biasing circuitry to restore equilibrium. When operating in this mode, the current increase due to the arrival of a photon is much more marked since the only current without the photon is the dark current.

The number of photons of a monochromatic light is given by the expression:

$$P = \frac{\lambda H}{hc} \text{ photons/m}^2/\text{sec} \qquad \text{Equation 14-1}$$

where λ = wavelength in meters, H = light flux in watts/m²/sec, h = Planck's constant (-6.625×10^{-34} watt-sec), and c = velocity of light in free space (3×10^8 m/sec).

If the diode is operated at something near the avalanche or zener voltage, a considerable amount of gain is obtained in terms of current/photon. However, this is another example of a case where nature seldom provides a free lunch. In the avalanche mode, the diode is subject to considerably more noise output that in the simple reverse biased mode. Therefore, the net signal-to-noise ratio may actually decrease.

The spectral response for the diode is controlled by a number of items. First, for the pn junction itself, photons must have enough energy to cross the band gap. This factor sets the ultimate low-frequency or long-wavelength limit upon the unit. However, a number of other factors affect the overall response as well. The photon energy limit would set only the lower limit of the response, and at shorter wavelengths, where the photon energy is higher, the photon efficiency might actually rise, giving a flat response or perhaps a rising one. However, this is *not* the case. Crystalline silicon is not particularly transparent in the visible range, and the photons usually have to penetrate some distance into the silicon or germanium in order to reach the junction. Also, there is nearly always some layer of glass which must be penetrated. This is usually quite transparent in the visible range and quite opaque in the ultraviolet. As a result, the spectral response of the typical photodiode is generally a peaked affair, as shown in Fig. 14-6A. This is a typical silicon photodiode response that has been just about optimized for the low attenuation pass bands of the optical fibers in the 0.8 to 0.9 μm range.

The bandwidth that is most often limiting to the communication system designer is not the spectral bandwidth but rather the electrical bandwidth of the diode. One of the limiting factors in the electrical

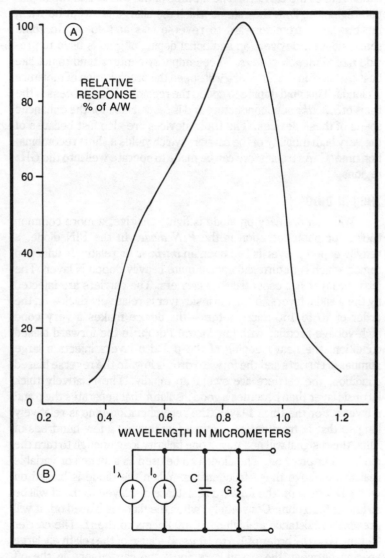

Fig. 14-6. Silicon diode spectral response at A, and the electrical equivalent circuit at B.

bandwidth is the capacitance (designated C) in the equivalent circuit in Fig. 14-6B. The photocurrent of the diode has to charge this capacitor to change the current in the external circuit. The conductance shown symbolizes the voltage drop across the diode.

A second factor that limits the speed of a diode is the recombination rate of the carriers. The lifetime of the carriers in the depletion region can limit the rate at which the charge is swept up when the bias goes from forward to reverse bias and so on. In some semiconductor devices, an additional doping of gold is used to provide *recombination centers*. These impurity centers tend to act like dust particles in the air which will speed the precipitation of moisture in clouds. This gold doping to speed the recombination process is the basis of *Schottky* semiconductors and is responsible for the enhanced speed of these devices. The GaAs devices are also fast because of the very high mobility of the carriers which yields a short recombination time. GaAs transistors can be made to operate well into the GHz regions.

THE PIN DIODE

While an ordinary pn diode is light sensitive, a more common choice for photodetection is the *PIN diode*. In the PIN diode, a heavily doped p layer is laid upon an *intrinsic* or relatively undoped region which is in turn laid upon a quite heavily doped N layer. The intrinsic layer has essentially no carriers. The carriers are injected by the p and n layers. If the intrinsic layer is relatively thick—on the order of 10 to 100 micrometers—the device makes a very good high-voltage rectifier with low forward drop. In the forward biased condition, the heavy doping of the p and n layers injects a large number of carriers and the forward drop is low. In the reverse biased condition, the carriers are swept up rapidly. The relatively thick intrinsic layer then becomes a good insulator that separates the p and n layers. For the thick I layer, the recombination time is relatively long so that at frequencies above a few tens or a few hundreds of Mhz, the rf signal will not stay in one polarity long enough to turn the diode either on or off. The diode can be used as a switch or variable resistance above these frequencies. When the diode is biased on with a DC current, the forward resistance displayed to the rf will be as low as 0.25 ohm. Conversely, when the diode is biased off, it will present a resistance as high as 10,000 ohms to the rf. The carrier lifetime is on the order of 1.5 to 10 μs. Because of the relatively large spacing between the p and n regions, the capacitance in the off condition is very small, typically less than 1 pF. Most of this is associated with the package rather than the junction itself.

On the other hand, PIN diodes with very thin I regions on the order of 5 micrometers or less have storage times which are measured in nanoseconds or less. These devices are useful, for instance, as microwave storage devices in parametric amplifiers.

In the photodetector application, the impact of the photon upon the I region can create hole-electron pairs which supply the photocurrent. Because the I region is a relatively good insulator, the dark current is low, which is advantageous in this application. The relatively small capacitance tends to speed the detector operation. The responsiveness of the PIN diode is also relatively good. For example, a Hewlett-Packard 5082-4207 PIN photodetector will demonstrate a responsivity of 0.5 amperes of photocurrent per watt of optical input at the peak of the response curve. This is coupled with a rise time of 1 nanosecond. This diode would typically be reverse biased at about 20 volts.

The avalanche photodetector is usually biased much higher, on the order of 200 volts reverse bias. In comparison, the RCA C30817 has a rise time of 2 nanoseconds and the responsiveness varies between 5 and 100 amperes per watt. This variation is due to the bias. If the avalanche diode is operated with a substantial margin between the zener voltage and the bias, the gain due to the avalanche effect is small. On the other hand, as the zener voltage is approached, the avalanche gain increases until it becomes theoretically infinite precisely at the avalanche point. Exactly at this point, a single photon would theoretically completely break down the diode and the current would continue to flow indefinitely.

In actual practice, the zener voltage is not constant. The temperature of the chip has a significant effect upon the zener voltage. The photocurrent and the light flux can both alter this temperature slightly. Therefore, there is a limit to the closeness with which the bias voltage can approach the avalanche voltage or zener voltage. For the RCA detector cited, the unit can be run with a tightly controlled and tracked bias at the level of 100 amperes per watt. However, a more practical level is 50 amperes per watt. At the reduced gain, the required temperature tracking is much less critical and the diode noise is smaller. In digital systems, the *avalanche photodetector* (APD) will typically demonstrate an overall sensitivity of 15 to 17 dB greater than a PIN diode amplifier system.

THE PHOTOTRANSISTOR

The basic structure of a transistor is shown in Fig. 14-7. Compare this with the pn junction shown in Fig. 14-4. The transistor shown is an npn type. However, the operation of the pnp is very

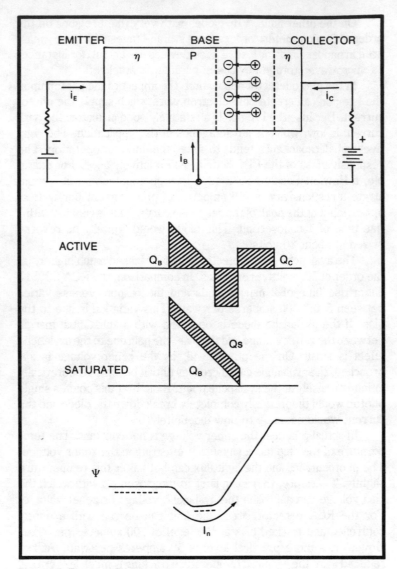

Fig. 14-7. The transistor.

similar. The transistor can often be considered as two diodes. The base-emitter junction forms one diode, and the base collector junction forms another. As shown in the circuit of Fig. 14-7, the base emitter junction is forward biased and the collector emitter junction is reverse biased.

For normal transistors, the doping levels are chosen so that the forward biased emitter junction conducts by injecting electrons into the base and only a tiny fraction is due to holes injected by the base into the emitter. The base is also constructed thin enough so that nearly all of the injected electrons diffuse to the edge of the depletion region in the reverse biased collector base junction. At this point, the field due to the bias sweeps them up into the collector bulk. The fraction of the emitter current reaching the collector is usually designated as α (Greek alpha). The remainder of the current, which is $(1-\alpha)i_E$, represents the holes being injected by the base into the emitter. This gives a ratio of collector to base current:

$$\beta = \frac{\alpha}{1-\alpha}$$ Equation 14-2

where β = current gain.

The transistor can be used to provide gain or amplification in a variety of ways. First of all, the device will exhibit current gain if the emitter is grounded—that is, the collector current will be much larger than the base current. In high gain small signal transistors, the value of β is more than 100, so the current is amplified more than 100 times.

The second mechanism for amplification stems from the fact that the impedance of the collector circuit is much higher than the impedance of the base circuit. With the control signal injected into either the emitter or the base circuit, a high voltage can be developed across the high-impedance collector load resistor. When the collector load is large enough with respect to the collector current, the load will limit the current and the device will be *saturated*. The collector field in this case will be sweeping up as many of the electrons as permitted by the external circuit, and the collector-emitter potential of the transistor will be minimum.

In a phototransistor, the base current is caused by the photon-generated hole-electron pairs in the base-emitter junction. The unit will experience the same gain in collector current as shown in the amplification case. The collector current will be given by:

$$i_c = (\beta + 1) i_\lambda$$ Equation 14-3

where i_λ is the photocurrent induced in the base emitter photodiode.

The phototransistor can be operated either with the base left open as regards the external circuit. Then only the amplified dark

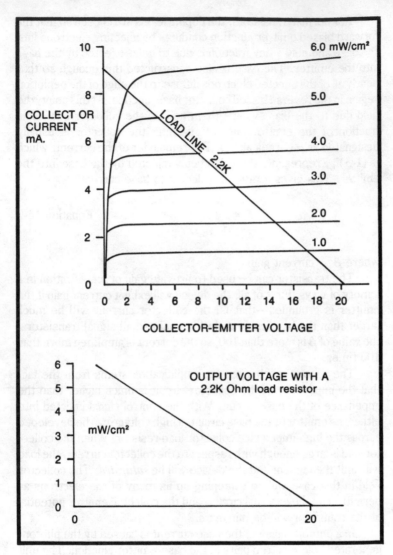

Fig. 14-8. Phototransistor response curves.

current will flow in the collector. The base emitter junction can be forward biased so that the phototransistor draws a current in darkness on the order of 1 to 10 mA. This has the effect of increasing β (actually small signal current gain h_{fe}) and thereby increasing the overall sensitivity of the detector.

Figure 14-8 shows the voltage current family of curves for a typical phototransistor, the Motorola MRD 300. The response in terms of current is reasonably linear in the 0 to 4 mw/cm² range. As the photocurrent rises into the 6 to 10 mA range, the increasing current gain of the transistor with current in this range comes into play and the device becomes a little nonlinear. On the whole, the curves cannot be distinguished from the normal curves taken as a function of base current. The linearity could be improved by shifting the base bias of the transistor so that with higher photocurrents, the collector current would stay below 6 mA.

By comparison with a photodiode, if a transistor were run in a constant collector to emitter voltage, between 2 and 3 mW/cm² the unit would have a responsiveness of 1.5 mA/mW/cm². If the transistor is operated as a voltage output device with the aid of a 2.2K ohm load resistor a response of nearly 4 volts per mW per cm² can be obtained. This response is actually very much larger than that of the diode because the diode response was quoted in amperes per watt. If the photodiode were to have an area of one square millimeter, an excitation of one milliwatt per square centimeter would input only 0.01 milliwatts or 10^{-5} watts into the diode. At a sensitivity of 0.5 A/W, this would correspond to a current of only 5 microamperes. For the APD operating at 50 A/W, this would correspond to 0.5 mA. In terms of actual sensitivity, the phototransistor is roughly as sensitive as the APD and is much easier to use.

That is the good news about the phototransistor. The bad news is that the phototransistor has a base emitter capacitance of something like 4 pF. This effectively tends to limit the rise and fall times to something like 2 to 3 microseconds. Effectively, the analog 3 dB down frequency is held to about 50 KHz. For slow-speed data, the phototransistor is ideal. For any really high-speed applications, however, the phototransistor cannot compete with the PIN diode or the APD.

For applications requiring still higher gain, Photodarlington transistors such as the Motorola MFOD300 exist. This unit provides 75 mA/mW/cm² and consists of a two-transistor amplifier: a phototransistor and an output transistor. The emitter of the phototransistor is internally connected to the base of the output transistor in the Darlington amplifier connection. The total current gain of the ensemble therefore becomes the product of the current gain of the phototransistor and the current gain of the output transistor. The output current is thus some 50 times higher than that of the phototransistor alone. Unfortunately, it is characteristic of the Darlington

259

amplifier that the effective base-emitter capacitance is also multiplied. For this reason, the rise time on the unit is something like 40 μs and the fall time is something like 60 μs. This device has an upper 3 dB point at something like 7 kHz, which is adequate for analog speech but a bit shy for analog music of even broadcast quality.

15

Light Sources

We noted previously that the incandescent lamp is not a very suitable source for obtaining light for a fiberoptic communication system. For endoscope or fiberscope usage, where the object is to present an image suitable for viewing with the human eye, the miniature incandescent lamp may still be the light source of choice. However, for communications systems, other requirements prevail.

LINE WIDTH

One of the prime characteristics desirable in a light source for communications work is narrow line width. In reality, the narrower the line width of the source, the better the system will work. The reason for this is the *intramodal dispersion* of the fibers, which is caused by the fact that the index of refraction is not precisely constant with respect to wavelength.

Referring back to Chapter 9, the index of refraction of crown glass is as shown in Table 15-1. If you take the difference in wavelength and the difference in index, you'll find that the rate of change of index (assuming a linear slope) is −48002/meter. Suppose that you had a source which completely filled the *window* shown for the

Table 15-1. Index of Refraction for Crown Glass.

	λ	η
C Line	656.5×10^9 meters	1.51263
F Line	486.3×10^9 meters	1.52080

example fibers between 800 and 900 nm. You could calculate the difference in index across the band as -0.0048; that is, the index at 900 nm would be lower by 0.0048 than it was at 800 nm. If you presume that a similar dispersion applies to fused silica in the fiber, you would have an index of something like 1.4600 at 800 nm and 1.4552 at 900 nm. Then the time required to propagate 1 Km at 800 nm is:

$$1000/3 \times 10^8 \times 1.46 = 4.86667 \times 10^{-6} \text{ sec}$$

and at 900 nm:

$$1000/3 \times 10^8 \times 1.4552 = 4.8500 \times 10^{-6} \text{ sec}$$

The difference in propagation time is 16×10^{-9} sec sec/km.

For an analog system the intramodal dispersion alone would limit the bandwidth to something like 22 MHz. If the optical source bandwidth were reduced to 10 nm, the signal bandwidth would be increased to 220 MHz/km.

The intramodal dispersion becomes most important for the graded index fibers. However, for a wide-bandwidth system, the intramodal dispersion can easily become a significant factor in the system rise time or bandwidth determination.

Of course, the light from a broader source could be filtered to eliminate all but a narrow band, but as shown in the monochrometer example discussed earlier, this is done only at the expense of wasting the light. Glass dye or interference filters can be obtained with a linewidth of only 2 angstroms or 20 nm. For the interference filters, the index of refraction is abruptly increased at intervals of an odd number of quarter wavelengths. In this condition, as discussed in the transmission line propagation section, the reflections due to the impedance mismatch will cancel only for the wavelengths where the intervals meet the quarter-wavelength condition. The filter can thus pass only a very narrow band of light. However, for these conditions the light rejected by the filter is simply lost to the system. A better solution to obtaining very narrow bands of light is available from gas discharge lamps sold commercially under the name *Osram lamps*. For these lamps, the light is confined to the spectral lines of the gas or vapor. In the case of multiple lines, some of the lines must be rejected and only a single line used. The loss is much less, however, than in the filtering of a single broad source, such as an incandescent lamp.

BRIGHTNESS

In addition to having the light as nearly monochromatic as possible, it is desirable to have the source as bright as possible since the fiber can capture only a tiny area. A lamp such as a *neon* bulb has a reasonably narrow bandwidth, but the surface density of the light is very low. A relatively low density discharge is distributed over the area of the electrodes in the lamp. Only a very tiny fraction of this can be picked up by the end of the fiber. By comparison, the amount of light that is generated in a single tiny spot in a LED is many orders of magnitude brighter. The Osram lamps also suffer from this difficulty because it is difficult to obtain the required density of the electrical discharge in watts per cm^3 without destroying the lamp. Furthermore, the multiple envelope design of Osram lamps makes it difficult to get a fiber close enough to pick up any significant part. An Osram lamp usually has an outer envelope about 1 inch in diameter and perhaps 5 inches long. The actual discharge takes place in a glass vessel perhaps one-quarter inch in diameter and five-eighths-inch long, more or less centered within the outer envelope.

EASE OF MODULATION

For any high-speed communication, the light source itself must usually be capable of having its output easily varied or at least turned on and off with a rapid rise time and fall time. Even modest speed systems require changes that are far too fast for any mechanical shutter or iris arrangement. An incandescent lamp of any size simply heats and cools too slowly to permit modulation beyond about a kilohertz. The gas discharge lamps will do better than this, but their rise and fall times are still measured in tens or hundreds of microseconds. A few electrooptic devices, such as the *Kerr cell,* can achieve suitable speeds, but these usually require considerable amounts of power or very high voltages to drive. There is a very great practical advantage to having a source that can be modulated by a simple variation in the drive current or voltage with rise and fall times of a few nanoseconds or a microsecond.

OTHER ATTRIBUTES

With the continuing trend to miniaturize electronic gear, there is an obvious advantage to having a source which is physically tiny. The control box for an Osram lamp typically is about a 9-inch cube and can weigh 10 to 20 pounds with necessary transformers and controls. This is obviously not a very attractive prospect for a miniature electronic system. Also, the power consumption is on the

Table 15-2. Emission Range for Various Materials.

GaP	Green	560 nm	665 lm/W
GaAsP/GaP	Yellow	580 nm	570 lm/W
GaAsP/GaP	Light Red	635 nm	(High efficiency red) 147 lm/W
GaAsP	dark red	655 nm	(standard red) 55 lm/W
GaAs	infra-red	900 nm	(invisible)

order of 150 watts, which can pose a problem in certain applications.

Any engineering project is optimized when the cost and size are minimal and all of the performance goals are met. The relatively low cost of a tiny LED source operating on less than 1 watt is therefore attractive. However, a number of costs enter into a workable fiber-optics link, and the source constitutes only a portion of the cost.

THE LED

Probably very few people will read this book who are not by now familiar with *light-emitting diodes* (LEDs) or at least the sight of them in calculators, watches, etc. The use of these devices has grown rapidly since their introduction because of the relatively low power consumption compared to their brightness and small size.

Whenever a forward current is passed through a pn junction, a certain amount of recombination of the excess holes and electrons takes place. During this recombination, a quantum of energy must be given up. This quantum of energy is roughly equal to the band gap energy or slightly greater due to thermal effects. The energy can be given up in the form of photons of visible energy, or by *phonons* (a mechanical vibration or sound in the lattice). The wavelengths are thus roughly determined by the nature of the crystal structure or the band gap in the structure. Thus the color of the emitted light is dependent upon the nature of the material. For the various materials, the emission range is approximately shown in Table 15-2.

Note again that the lumens per watt refer to the radiation perceived by the standard photropic eye. In communications work, we are more interested in the number of watts of light put into the fiber.

Figure 15-1 shows the forward drop of a number of diodes with a range of emission wavelengths plotted against the forward voltage drop for a 10-mA forward current at a junction temperature of 25° C. The forward voltage drop is roughly proportional to the wavelength as you might expect and is related to the reciprocal of wavelength. The dotted curve has been empirically fitted to the data and shows that the distribution is not really linear with the reciprocal of wavelength but rather varies as the reciprocal of wavelength to the 1.675 power for the data fitted.

Data of this sort are relatively difficult to curve fit with any great accuracy. For one thing, the forward drop of the diodes is influenced to a large extent by the construction of the diode and the nature of the ohmic contacts made to the p+ and n− regions of the diode. In addition to this, the forward current is strongly influenced by temperature and the voltage current curve is very nonlinear. This all

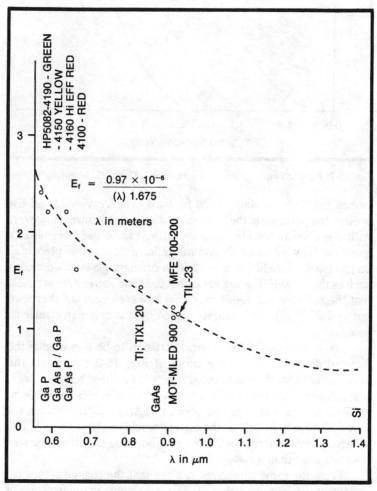

Fig. 15-1. Forward drop for various LEDs. I is 0.01 amperes, and t is 25° C.

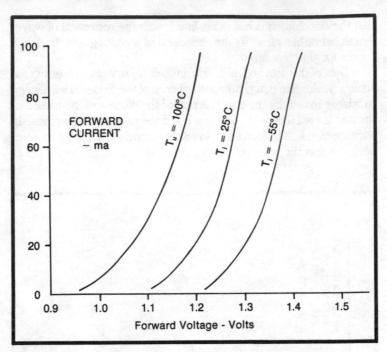

Fig. 15-2. Forward voltage versus current of a TIL-23 GaAs diode.

makes the determination of the forward drop very difficult if the precise temperature of the pn junction itself is not known. However, with these restrictions in mind, the curve can be used to answer a question: Why is it that if the recombination process takes place in all pn junctions, no radiation is seen from ordinary glass-cased diodes such as the 1N914? The answer is that the band gap energy is so low that the radiation emitted is very long in wavelength and therefore not visible. The 1.4 micrometer wavelength is only perceivable as heat.

As an example of the thermal variation to be expected in the forward drop of a LED, the curve of Fig. 15-2 represents the variation of forward voltage versus current for three temperatures. If a forward voltage of 1.25V were maintained across the diode at −55°C, the current would be about 8 mA. At 25°C, the current would be 45 mA. At 100°C, the current would be about 130 mA. It is relatively obvious from this that a LED should be driven by a *current* source rather than a *voltage* source.

From the same curve, you can see that the change in forward voltage at the 50-mA level is −0.2V for a change in temperature of

155°C or 1.29 mV/°C. This is somewhat less than the nominal 2mV/°C that is usually quoted for silicon pn junctions.

Figure 15-3 is also presented for the TIL-23 LED. It offers an interesting comparison. This diode is optimized for use with silicon phototransistors (or nearly). It shows the relative responses of the human eye and the phototransistor, as well as the light output curve for a tungsten lamp and the TIL-23. The spectral output of this source has a half-power bandwidth of 50 nm, which is respectably narrow for a LED. By way of comparison, the Motorola MLED 900 has a half-power bandwidth of 80 nm.

The TIL-23 is a metal cartridge-cased unit with a lens installed. It will operate at a maximum continuous current of 100 mA at a case temperature of 25°C. It is rated for pulse operation to 500 mA at a 10 percent duty cycle for a case temperature of 25°C. Under this condition, it will provide a 2.8 mW output at the peak. Under these conditions, assuming a forward drop of 1.75 volts, the peak power input would be 0.875 watts and the electrical to photon efficiency would be 0.3 percent. At lower, continuous currents the electrical efficiency will rise to 1.5 percent at 25°C.

By comparison, the Motorola MLED 900 is a smaller affair with a crystal-clear molded plastic case about the size of a "BB" shot. The flattened leads leave the molded plastic button radially, and a tiny

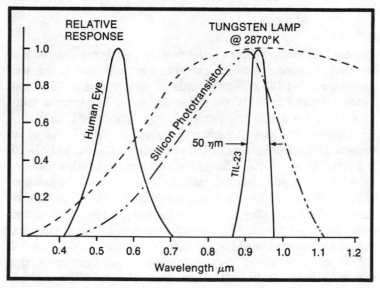

Fig. 15-3. LED output response curve.

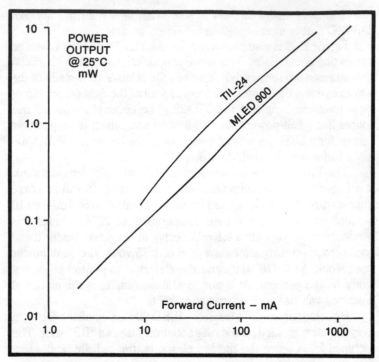

Fig. 15-4. Power output versus forward current for the TIL-24 and MLED 900.

rounded lens forms the front of the device. This device is rated at a maximum forward continuous current of 80 mA. The device will dissipate a rated 120 mW at an ambient temperature of 25°C and must be derated 2mW/°C for higher ambients. The junction temperature of this device has a thermal resistance of 500 °C/W.

Consider that you would like to operate the MLED 900 at an ambient temperature of 50°C. This would derate the unit to 120−25 × 2 mW/°C, or 70 mW. The maximum allowable junction temperature is 85°C. If you wished to limit the temperature of the junction to 70°C, this would allow a maximum of 20/500°C/W, or 40 mW dissipation. This reading is less than the temperature derating and would therefore control. At a forward voltage of 1.2 volts, a current of 33 mA continuous would be the limit. The diode would output 0.4 mW at 25°C. However, as the junction temperature rises, the output of the diode falls. The rise from 25° to 70°C would be accompanied by a drop in output by a factor of 0.65. Thus the final

268

output of the diode is 0.26 mW as the junction temperature reaches 70°C.

There are differences in linearity for the devices. The TIL-23 and TIL-24 are essentially identical, with the TIL-24 being the premium part. The curve of power output versus forward current for the TIL-24 and the MLED 900 are shown in Fig. 15-4. The curve for the TIL 23 would cross the MLED 900 curve at 10 mA and be parallel to the curve for the TIL-24. Below about 40 mA, the curve for the TIL units is nonlinear. These units are primarily intended for pulse operation where this nonlinearity would be no disadvantage and the higher output would be an advantage. On the other hand, the superior linearity of the MLED 900 would be an advantage if the system were to be used for *analog* operation. The poor linearity of the TIL units would cause a substantial amount of intermodulation distortion if the units were operated at a quiescent bias on the order of 50 mA. In a pulsed application, however, the advantage goes clearly to the TIL units. It should be noted that the units are not really directly comparable. The TIL units are relatively expensive units in metal cases with silica lenses, whereas the MLED unit is in a relatively inexpensive molded plastic case.

Also, the peak of the spectral line shifts as a function of the junction temperature. In rising from 25°C to 75°C, the TIL units are rated to increase their output wavelength by 50 nm. The coefficient is rated at 3 nm/°C.

Both of the types are lensed and have a definite focused beam output. For the TIL units, the half-power beamwidth is 35°. For the MLED, the half-power beamwidth is 28°. This means that the power has dropped to the 50 percent level 17.5° and 14° off-axis, respectively. However, this figure is not necessarily complete because this is the far-field beamwidth. Figure 15-5 shows this problem. Since the actual LED chip inside the package has a finite size, some offset in the image of the chip is focused at infinity. In fact, for most lensed LEDs the fact that the half-power beamwidth is not very tiny is due to the short focal length of the lens and the finite extent of the LED chip itself.

For example, the half-power beamwidth of a 2-mm lens operating at 0.9×10^{-6} would be given by:

$$1.14 = \times 9 \times 10^{-7}/2 \times 10^{-3} = 5.13 \times 10^{-4} \text{ radians}$$

This equals 0.024 degrees half-power beamwidth. This is clearly a tiny fraction of the typical beamwidths of lensed LEDs. On the other

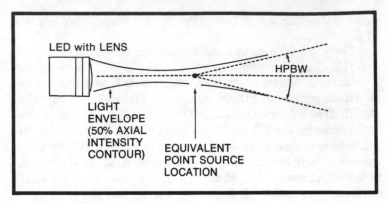

LED with LENS

HPBW

LIGHT
ENVELOPE
(50% AXIAL
INTENSITY
CONTOUR)

EQUIVALENT
POINT SOURCE
LOCATION

Fig. 15-5. The beam effect of a lensed LED.

hand, if an LED has a chip size of 1.4 mm with a lens that has a focal length of 3 mm, the center to the corner angle of the system is 14 degrees and the lens could focus the light no sharper.

A second factor comes from the fact that even if the LED were a true point source, it would still fill the lens with light. Only at some distance from the lens could the beam be described in terms of angles. Figure 15-5 shows this effect. The far-field acts like a point source located at some point outside of the lens. There is always some optimum point slightly separated from the lens at which to place the coupling fiber. This point can be located in one of several ways; it can be found by a careful probing of the field in front of the lens with a three-axis positioner or it can be located by using a visible LED which is mechanically and beam identical to the IR source eventually to be used. Identical sources in the visible range are not offered for all sources; however, Fairchild offers the FLV 104 as a companion in the visible range to the FPE 104 narrow-beam LED transmitter. In most cases, however, the three-axis probing is probably required to produce the best coupling between a lensed LED and a given fiber.

The physical construction of a LED for fiberoptic use presents some significant problems. There are obviously differences between the requirements for a LED to be used for a signal lamp or a part of a display and a diode which is intended to be used for a fiberoptic source. The first and foremost among these is the matter of power handling. Most visible signal LEDs will handle something on the order of 10 mW input for the smallest sizes and 100 mW for the larger sizes (similar to a TO-18 transistor size). On the other hand, the light source for a fiberoptic system should be capable of producing the largest possible amount of light if it is to be used in a long system since the higher the power, the longer the permitted distance between repeaters. Accordingly, the sources for fiberoptic usage should have good heat sink systems.

A second requirement is the ability to get the fiber as well coupled to the active area of the light source as possible. In a visible signal lamp, a relatively large active area is not a particular drawback. In the fiberoptic source, any energy generated in an area outside of the fiber proper is usually lost. A lens system can help; however, as shown earlier, it does not really do the trick as well as you might hope.

Figure 15-6 shows the construction of a typical TO-18-sized LED for signal purposes. The chip is mounted upon a relatively large pin which serves as the cathode lead for the diode. The p junction is deposited upon the top. An ohmic contact in the form of a dot is placed in the center of the chip. A gold wire is bonded to the dot and welded to the cathode lead. The entire top of the chip glows, except for the dot in the center. The heavy cathode lead serves as a heat sink for the chip.

For this construction, the flat top of the chip serves as a Lambert's law radiator, and the diode would be visible over a very wide angle. The inclusion of a lens in the plastic package tends to brighten the image along the axis of the system at the expense of the width of the viewing angle. The dot in the center does *not* assist in the coupling of the light into a fiber.

To eliminate some of these problems, C. A. Burrus of Bell Labs came up with a certain design in 1971. The lower illustration in the figure shows the construction which has become known as a *Burrus diode*. The chip is bonded directly to a heavy heat sink stud which carries the heat directly out of the chip to the case. The chip constitutes a high-radiance LED. A well is etched through the top contact layer to expose a small active area. The metal contact for the anode is also relatively heavy and serves as a heat sink.

The fiber can be placed in the Burrus diode right up against the active area and there is therefore very little spreading loss. The active area is a Lambert's law radiator. Consequently, some radiation is lost due to the inability of the fiber to accept wide-angle radiation. For fibers with a small NA, this loss can be substantial. However, the closeness with which the fiber can approach the source still makes for a very high coupling, and LEDs with outputs of 1 mW from the fiber pigtail are commercially available. In general,

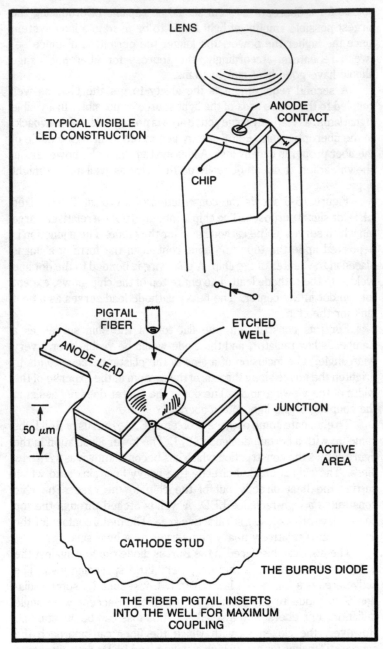

LENS

TYPICAL VISIBLE
LED CONSTRUCTION

ANODE
CONTACT

CHIP

PIGTAIL
FIBER

ETCHED
WELL

ANODE LEAD

50 μm

JUNCTION

ACTIVE
AREA

CATHODE STUD

THE BURRUS DIODE

THE FIBER PIGTAIL INSERTS
INTO THE WELL FOR MAXIMUM
COUPLING

Fig. 15-6. Construction of a typical TO-18 sized LED used for signal purposes.

the diode is purchased with the pigtail fiber installed as a permanent part of the diode package. The diameter of the well in the Burrus diode runs typically between 50 and 75 μm, so little light will be lost on fibers of these diameters.

A second popular geometry is the *stripe contact edge emitting source* shown in Fig. 15-7. In this configuration, the metal contact is deformed to present a ridge or stripe running the width of the diode. Because of this ridge, the field is highest along the ridge area, so most of the diode conduction takes place in this area, even though the junction extends through the entire chip. The active area is shown hatched in the figure. This active area can have a height of from 1 to 25 μm and a width of 20 to 150 μm. The light issues from the edge of the diode chip. Half of the light will of course be lost out of the backside of the chip. In some newer designs, however, a photosensor is placed in the rear to monitor the diode output and adjust

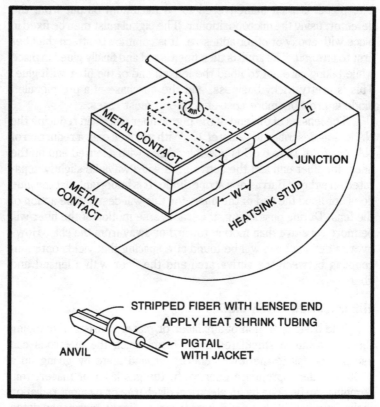

Fig. 15-7. The stripe edge emitter diode.

the drive accordingly. The stripe configuration is also used in the injection laser diode.

With a stripe edge emitter, the unit can couple fairly effectively to relatively small fibers, but the NA matching loss is still present in a LED, as with the Burrus diode. For single mode fibers, a considerable area matching problem can still exist. For example, suppose that the fiber has a 2.5 μm core and the LED has a 1×25 μm active area. The fiber will cover only about one-tenth the active area, so nine-tenths of the light will completely miss the fiber for an area coupling loss of 10 dB. On multimode fibers, the core would be larger than the active area and there need be no area coupling loss if the fiber and the active area were accurately aligned.

The stripe edge emitters can be purchased from a number of sources with the pigtail installed. If you want to install the pigtail yourself, it is best done using a three-axis micropositioner and a photodetector attached to or spliced temporarily onto the pigtail. The pigtail is then maneuvered for a maximum output from the detector, using the micropositioner. The pigtail must then be fixed in place with epoxy or other adhesive. It is common to attach the fiber first to an *anvil*. The anvil is then positioned and finally glued in place while taking care not to flood the lensed end of the fiber with glue. This is pretty tricky business, and the purchase of a pre-pigtailed diode is probably more cost-effective in most cases.

The lensing of the end of the pigtail fiber will aid in reducing the NA loss as described in Chapter 12. With a lensed end, a reduction of NA loss from 12 to 6 dB is possible. Also, with a lensed end on the fiber, the fiber end and the active area will always be slightly separated on either a Burrus or a stripe diode. It is important that the glue does not flood the end of the fiber, for this will destroy the action of the lens. During positioning, the transverse motion of the fiber will be more sensitive than motion toward or away from the chip. However, a definite peak will be found at a spacing that yields optimum coupling between the active area and the fiber with a lensed end fiber.

THE INJECTION LASER DIODE

The basic operation of any laser or maser consists of pumping the atoms into a stimulated state from which the electrons can escape and fall to the lower energy ground state by giving up a photon of the appropriate energy. In the gas laser or maser, this pumping can be done by an electrical discharge or by input radiation in the form of light. In a solid-state laser, the input energy populates some of the unpopulated bands. When a photon of energy equal to the band gap of the crystal passes through, it stimulates other excited photons to fall in step with it. Thus a small priming signal will emerge with other photons that have added coherently. Amplification will result.

The laser can be operated in a continuous wave (CW) oscillator mode if the ends of the laser path are optically flat and parallel and reflective to form a resonant cavity. A photon that is emitted from spontaneous falling of an electron through will rattle back and forth between the mirrors, acquiring companions due to the stimulated emission. In something like a Ruby laser, this stimulated emission can be so rapid as to represent nearly an avalanche, completely draining the high-energy states in a subnanosecond burst of very great energy. On the other hand, certain types of lasers can be produced which will operate in a more or less equilibrium condition, giving up photons at just the input energy pumping rate. The *injection laser diode* (ILD) will operate in this mode.

The material in an ILD is rather heavily doped so that under forward bias, the region near the junction has a very high concentration of holes and electrons. This produces the *inversion* condition with a large population of electrons in a high-energy band and a large population of holes in the low-energy bands. In this condition, the stimulated emission of photons can overcome the photon absorbtion and a net light flux will result.

A very common geometry for the laser diode is the stripe diode configuration shown in Fig. 15-7. The chip is about 0.010 inches on a side and may be made of AlGaAs for infrared operation with a peak emission at 0.84 μm. The ends of the stripe are prepared by cleaving to provide an optically flat and optically parallel set of mirrors. The index of refraction of the material is so high that the ends make effective mirrors. This parallel mirror structure is known as a *Fabry Perot interferometer* which can be constructed on a more macroscopic scale by carefully aligning two first-surface glass mirrors.

An initial thermally or spontaneously excited photon propagates down the stripe and stimulates the emission of other photons. The reflections at the ends keep a supply of photons rattling back and forth at all times, and the stripe acts as a resonant cavity. The cavity resonates only at the frequencies where the length is precisely equal to an integral number of half wavelengths; however, the optical length of the cavity is so great—900 wavelengths or more—that only a tiny shift in frequency or a tiny angle off-axis is required for an adjustment to another resonant condition. In the solid, there is always some thermal agitation of the lattice which tends to broaden the energy bands. Therefore, photons suitable for any given mode of resonance are available for stimulated emission, and the ILD will tend to operate in multiple cavity modes simultaneously.

For example, assume that the cavity is precisely 900 wavelengths long at 0.84 nm. A prime resonance would then take place at 1800 half-wavelengths, and a second resonance would take place at 1801 half-wavelengths, and so on. For this second mode, the wavelength would be 0.83953 nm or a difference of 4.66×10^{-10} meters. You would probably expect to see a spectral line every 0.466 nm. Such small differences in wavelength are difficult to resolve. In practice, the ILD will have a spectral width measured between the half-power points of 2 to 4 nm, which probably represents 5 to 10 modes.

In operation the forward current must reach some threshold to achieve lasing operation. Just beyond this threshold, the laser operates with a single "thread" of light, and the output is relatively stable but low. As the current is increased, the light output climbs very rapidly. The curve of Fig. 15-8 illustrates this action.

Several points in the curve are noteworthy. First, the rapidly rising curve is not terribly straight or linear. The curve is slightly exaggerated to bring home this fact of life. The ILD does not tend to be very linear for analog modulation. The Bell technique uses a photodiode behind the die to monitor the output and generate a feedback voltage to forcibly linearize the signal. The second point is that the light output curve is highly temperature sensitive. A rather modest drop in temperature can cause the output to rise so rapidly that the output can destroy the chip. The light monitor feedback can help in this area as well. Bell instead supplies a heater in a closed loop servo to hold the heat sink temperature constant. The voltage versus current curves of the ILD are nearly identical to those of a LED, but they cannot be reproduced from unit to unit. Like the LED, the ILD must be driven from a current source rather than a voltage source.

To put the bandwidth of the ILD in perspective, a quartz crystal oscillator operating at 1 MHz, for example, will have a typical noise bandwidth of less than 1 Hz at the -3 dB points so the linewidth will be about 1 part in 10^6 or better. By comparison, the ILD will have a noise bandwidth of 2 to 4 in 840 or 1 in 420 or 210. In addition, the exact frequency of the peak output is prone to shift with temperature and drive current. Translated into frequency bandwidth, the noise band of the ILD represents 8×10^{12} Hz, or 1.6×10^{13} Hz. This means that the information, even with a narrow-band ILD, is being carried upon a carrier with a noise bandwidth much wider than the information bandwidth. For this reason it is not practical to consider such techniques as frequency modulation or single sideband modulation that require true coherent detection in which the actual frequency difference between the received carrier and a target or average carrier are measured. The ILD can be used only in *simple amplitude modulation* (with due adjustment for linearity) and in *pulse modulation* where the ILD is cycled on and off to convey the code.

On the other hand, the light emerging from the active area of the ILD is sufficiently coherent to provide a narrow beam whose width is determined largely by the dimensions of the active area. Typically with the ILD, the junction is about 1 μm in thickness. At a wavelength of 0.8 μm, this implies that the minimum beamwidth is something on the order of 40 degrees. Across the width of the stripe, the active area measures something like 20 μm. The radiation is not diffraction limited in this plane. A diffraction-limited active area of this width would produce a beam of approximately 2.4 degrees width at the half-power points. Instead, the ILD will typically show a beam on the order of 10 degrees. This implies that the entire active area is not fully coherent and in phase. The odd elliptical beam of course provides a certain amount of NA loss in coupling. For the 10-degree plane (in the wide dimension of the stripe), the

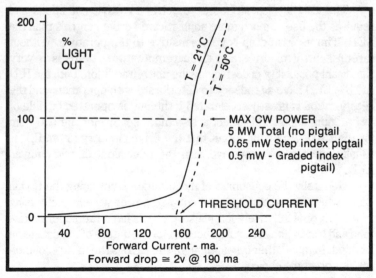

Fig. 15-8. ILD light versus current curves.

apparent NA is 0.087. Because this is smaller than the NA of most fibers, little loss is in this plane. In the narrow plane which gives rise to the wide dimension of the beam, however, the 40-degree half-power beamwidth corresponds to an NA of 0.342, which is wider than a number of fibers. However, the collimated beam of the ILD still clearly has the advantage over the Lambert's law distribution of LEDs. And the coupling is clearly superior.

One of the advantages of the ILD is that the residual reactances and shunt capacitance of the ILD are small and the lasing threshold is relatively sharp. If the diode is held at a current just below the lasing threshold and then rapidly raised above the threshold, extremely fast rise times on the order of 1 ns can be obtained. In very short pulses—on the order of 10 ns width with the diode cooled to liquid nitrogen temperature—pulses as large as 1 kW have been obtained. With the same cooling, CW powers of 3 W and pulses as narrow as 0.2 ns have also been achieved. In terms of performance, the ILD is clearly the star performer; however, it is also the most expensive. A pigtailed ILD runs costs about $650, and a pigtail stripe LED of about the same construction and power is similarly priced. Lensed LEDs can be obtained at prices from $50 down to the cheapest IR units packaged in molded plastic for perhaps $2.

The ILD is also far more "touchy" to drive conditions and requires the use of much more sophisticated drive controls than the LED. The forward drop is very sensitive to temperature. Without careful control of drive and diode temperature, there is a very significant possibility of destroying the unit. In addition, both the ILD and the LED have a tendency to deteriorate with operation, and the deterioration is greatly accelerated if the unit is operated outside of its optimum limits. This effect is also more severe for the ILD than for the LED. And don't forget that the drive circuitry for an ILD is usually much more expensive than the equivalent circuitry for an LED.

Generally the economics of the situation favor using the ILD in cases where the superior output coupled power will eliminate using repeaters on a relatively long link or else where the superior rise time and bandwidth is required because of the nature of the data to be handled. In most other cases, though, the LED is the more economical and practical choice.

WORDS OF CAUTION

Before leaving this chapter, several significant notes of caution should be stated. The first and foremost of these is a safety warning: Never look directly into either an ILD or an LED! Even for invisible output devices, the energy density is usually sufficient to cause eye damage. For the invisible units, the damage is all the more insidious because you do not *see* it. A burned spot can be produced on the retina before the victim is aware of anything! This applies equally to looking into the end of an active fiber. Here the coupling losses and the fiber attenuation reduce the hazard somewhat, but the risk of permanent damage to the eye is still substantial.

Never connect the device to a voltage source without some mechanism to limit the current. For voltages beyond the forward drop threshold, the current increases very rapidly. Even if a very precisely regulated voltage source is used, the unit can go into thermal runaway. As the device begins to draw current, the junction begins to warm and the forward drop decreases, which tends to drive the voltage higher and in turn decrease the forward drop until the current tends toward infinity and the device is destroyed. In testing a unit, it is always better to use a voltage source that is many times as large as the forward drop. A resistor which will control the current to a suitable value.

Always install the diode in the forward direction. Most of the LED or ILD sources have a reverse breakdown voltage of only about twice the forward drop, or on the order of 2 to 3 volts. They will avalanche or zener beyond this and draw very large currents if permitted by the supply circuitry. This is also a very simple and quick way of destroying a rather expensive device.

16

Care and Feeding of Light Sources

As noted in Chapter 15, the drive to the transmitter should be of a controlled current nature, rather than of a controlled voltage nature because of the steep voltage-current curves of the devices. There are also certain requirements due to the nature of the system. Analog systems must retain a high degree of linearity if distortion and harmonic generation and intermodulation product generation is to be minimized. The attainment of required system bandwidths can also prove to be problematical. This chapter will treat some of these problems.

To begin with, consider the attainment of a current control drive. In a mathematical example, suppose that you have a supply voltage of 5V and wish to drive the TIL-23 diode of Fig. 15-2 at a current of 80 mA. You can see from the figure that the forward drop for the diode is 1.29 volts at 25°C. This leaves 3.71 volts for control. If you elect to use a resistor for the current control, the resistor would have a value of 3.71V/0.08A, or 46.38 ohms. Next suppose that the diode temperature rose to 100°C. At 80 mA, the forward drop is 1.19 volts and the current would rise to 82 mA. The drop in forward voltage produces a higher current which in turn produces a slight increase in forward current.

As a second example, suppose that the source voltage is 10V. In this case, at 25°C the subtraction of the forward drop would leave 8.71 volts, and a control resistor of 108.9 ohms would give a current of 80 mA. The rise in temperature would give a rise in current to only 81 mA, half the rise shown with the 5V supply and smaller resistor. In general, the smaller a fraction of the forward supply voltage that is represented by the forward drop, the smaller the temperature effects that the change in forward drop has upon the unit current. Note that if a constant 1.19-volt source had been applied across the diode to produce a current of 80 mA at 100°C, the current would fall to 18 mA at 25°C. The object in driving these devices is to control the current and let the forward drop fall where it will.

It is not unusual in pulsed systems to use a resistor for current control in this manner. The active device is often a transistor used in the switching mode. A sample of this circuit is shown in Fig. 16-1. Even such a simple circuit requires a certain amount of design. The MLED 900 will withstand a continuous forward current of 80 mA and will dissipate 125 mW at 25°C. The power dissipation is to be derated by 2 mW/°C. Therefore, if 50°C operation is required, the dissipation should be held below 25 mW. With a forward voltage of 1.22 volts at 80 mA, this represents an average current of 0.02A so that the unit could have a maximum duty factor of .02/.08, or 25 percent. The thermal resistance of the device from junction to ambient is 500°C/W. The junction will therefore be 12.5°C above the ambient at 0.025 W dissipation. The ambient range would therefore be up to 37.5°C, or just above body heat. A more conservative design would be fairly restrictive with this device.

The 2N2222 (metal package TO-18) is rated to handle the switching of up to 500 mA and shows a maximum collector-emitter saturated drop of 0.3V at 150 mA collector current. Adding the drop for the transistor and the diode, 1.52 volts is obtained. Therefore, the drop across the current limiting resistor is 8.48V, and the 95 ohm resistor will control the current to 0.08A. The transistor has an h_{fe}, or *forward transfer ratio*, of 50 minimum, so a base current of about 1.6 mA is needed to obtain the collector current saturation. Actually, some margin is desirable. If you assume a base emitter drop of 0.6V, the 1.5K ohm resistor will provide 2.27 mA of base current. This should be adequate for drive. The unit will then present 1.5 standard TTL loads at the input.

The drive circuit will work well with a good grade of TTL. However, if the LO voltage gets up around 0.8 volts, the transistor may not turn off. The preferred circuit shown in Fig. 16-1 has the advantage of turning off more solidly even with marginally high L_0 voltages. It has the *disadvantage* of increasing the loading to 3 mA or two TTL standard loads.

The capacitance of this diode is rather large—150 pF—so the rise time with the resistor in place will be something in excess of 95 × 150 × 10^{-12} × 2.3, or 33 ns.

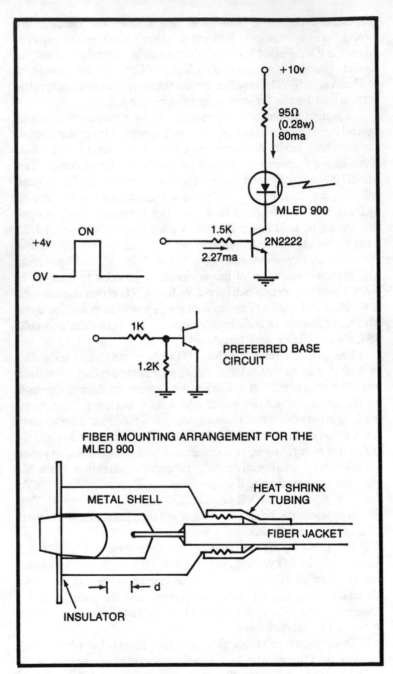

Fig. 16-1. A simple pulsed switch drive.

The allowance of about 2.3 × RC for rise time almost accounts for the specification of rise time from the 10 percent to 90 percent points. This is slower than the rise time specification for the transistor, which is a few nanoseconds and would therefore control. Probably the fastest pulse you would choose to use with this arrangement would have a 33-ns rise, a 99-ns dwell, and a 33-ns fall. For a 25 percent duty cycle, the unit would have to stay off for about 396 ns for a total pulse repetition rate of about 1.89 MHz.

The arrangement shown to the bottom Fig. 16-1 shows a mechanical arrangement for coupling the fiber to the diode. This unit is a very inexpensive LED in an all plastic package. The mount consists of a turned aluminum case which fits snugly on the diode and serves to center the fiber. The diode may be epoxied in place (carefully to avoid getting epoxy on the lens). The sleeve is bored to accept the clad fiber, and a length is bored to accept the fiber jacket. The piece centers the fiber before the lens of the LED, and the fiber is attached by means of the heat shrink tubing over the jacket of the fiber. The end of the fiber should preferably be lensed and the spacing d selected for maximum coupling to the fiber as shown by a maximum reading on the output. If the bushing or mount is carefully anodized, the leads from the diode can be pressed against the housing for good heat sinking without short circuiting through the anodized film.

A somewhat better choice for a low performance system might be the Motorola MFOE200 diode. This unit is in a metal TO-18 size package with a lens on the cap. The current ratings are nearly identical. However, due to the superior heat sinking, the unit is rated at 250 mW with a 2.5 mW/°C derating. The thermal resistance is

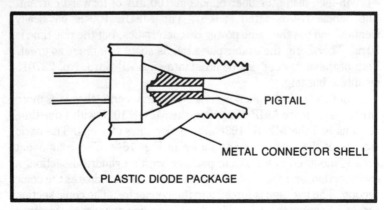

PIGTAIL

METAL CONNECTOR SHELL

PLASTIC DIODE PACKAGE

Fig. 16-2. The AMP connector.

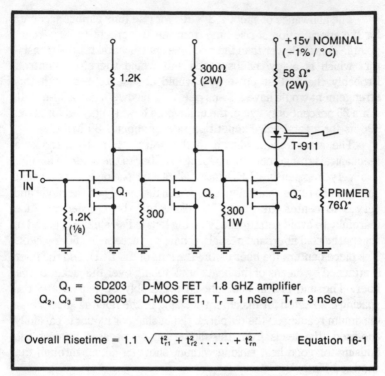

Fig. 16-3. A high-speed ILD driver for pulse circuits.

Inside the figure:

+15v NOMINAL
(−1% / °C)

1.2K

300Ω
(2W)

58 Ω*
(2W)

T-911

TTL
IN

Q₁ — Q_1

1.2K
(⅛)

300

Q₂ — Q_2

300
1W

Q₃ — Q_3

PRIMER
76Ω*

Q_1 = SD203   D-MOS FET   1.8 GHZ amplifier
Q_2, Q_3 = SD205   D-MOS FET, T_r = 1 nSec T_f = 3 nSec

$$\text{Overall Risetime} = 1.1 \sqrt{t_{r1}^2 + t_{r2}^2 \ldots + t_{rn}^2}$$ Equation 16-1

also lower at 400°C/W. This diode derates to zero output at 100°C compared to 60°C for the MLED 900. This is a rather slow unit with a rise time for the optical power of 250 ns, but it will deliver 29 μW into a 0.045-inch diameter fiber bundle at 100 mA of forward current. The bundle NA is rated at 0.67. The MFOE 100 is physically identical and has the same power characteristics, but the rise time is 50 ns. Therefore, the usable pulse rate is about five times as great. Both of these devices can be fitted into an AMP Corp. No. 227015 mounting bushing.

Motorola also makes a series of components that is a more direct usage of the AMP connector: the MFOE 102F with a rise time of 25 ns and the MFOE 103F with a rise time of 15 ns. The basic aspects of the connector are shown in Fig. 16-2. The component itself is mounted in a resilient package with a cylindrical section, a conic section, and a small cylindrical section that terminates the conic section. The package is forced into the connector. The conic section serves to center the assembly. A similar conic-cylindrical structure is used on the end of the cables, and a nut compresses the cable end into the connector. This connector ferrule is designated as AMP No. 227240-1. For these diodes, the output pigtail is a 200-μm core with an NA equal to 0.68. Motorola also supplies PIN diodes, phototransistors, and photoDarlingtons with the same mechanical configurations for the AMP connectors.

There are a number of other connector types. However, the AMP series seem to have a great many advantages for diode mounting.

ILD DRIVING FOR PULSE CIRCUITS

The ILD drive circuits are basically similar to the drives for the LED, but the ILD is generally not selected unless very fast pulses are required. In this instance, the diode must be *primed* to obtain maximum speed. Figure 16-3 shows a typical simple drive circuit for a pulsed ILD operation. The ILD is an ITT type T-911 AlGaAs room-temperature ILD. The construction of this stripe laser was shown earlier. The unit is rated for operation from 0° to 50°C and has a lasing threshold of 160 mA. The unit has an output wavelength of 0.84 μm with a 4-nm spectral width. The unit itself exhibits a rise and fall time of 1.5 ns with priming. At a current 30 mA above the lasing threshold, the unit has an output of 5 mW, of which 0.65 mW will be coupled into the pigtail that consists of an ITT type T-102 fiber. The fiber has a 125-μm core and is a low-loss stepped index fiber with an NA of 0.25 and a plastic outer jacket.

The transistors used for the driving proper are Signetics DMOS power FETs with fast rise times. These devices are cut off with a forward bias less than about 1.8V. With a forward or positive bias of about 5V, transistor Q3 will saturate and go to a forward drop of about 2V, thus shorting the primer resistor. The combination of the forward drop on Q3 and the forward drop on the ILD leaves about 11 volts, which is controlled to 190 mA by the resistor. The real object is to hold the current to 30 mA above the lasing threshold and in *no* case should the light output from the diode exceed 10 mW. The reason for the asterisk is that the resistor will have to be selected for the particular diode in use. In addition to this, the 160-mA lasing threshold has a temperature coefficient of 1 percent/°C, which is really moving pretty fast. If at turn-on, the threshold is 160 mA, the threshold will fall to as low as 136 mA as the temperature rises to 40°C. Obviously the heat sink would either have to be temperature controlled or else the resistor made variable with the temperature of the junction. The ILD is a $650 device so you wouldn't be anxious to burn out too many of them.

When Q3 is cutoff because of a gate voltage less than 1.8V, the current in the primer resistor will flow. In the nonlasing mode at 100 mA, the ILD will have a forward drop of something on the order of 1.6V, and the laser will generate a light output of about 3 percent of the peak output. The primer works in several ways. First, it keeps the lasing action just barely percolating. Second, it is not necessary for the diode current to swing so far, thereby reducing inductive effects.

Input transistor Q1 is another Signetics DMOS FET that is an ultrafast single-gate enhancement-mode unit which will exhibit gains of 10 dB at 1.8 GHz. This unit serves to drive Q2. If the circuit is properly laid out in microwave stripline fashion Q1 will exhibit a rise time on the order of 0.3 ns, and power FET Q2 will exhibit a rise time on the order of 1.5 ns and a fall time of 3 ns. Q3 will behave about the same and the ILD will show a rise time to optical output of less than 1.5 ns. Equation 16-1 of Fig. 16-3 shows that the overall rise time of a system is approximately given by the sum of the squares of the individual rise times taken to the one-half power. This calculation yields a rise time from the TTL level input of 2.88 ns and a fall time of 4.72 ns. If the unit stayed on for 10 ns and had a 50 percent duty cycle, the rep rate of the pulse train would be 28.4 MHz. To put this in perspective, it should be born in mind that in free space light travels 3×10^8 m/s. Therefore, in 1 ns light will travel only 30 cm or 11.8 inches. Every inch of the conductor between the ILD and the transistor will add one-tenth of a nanosecond to the rise time. In actual practice, the printed circuit board will have an index of refraction of something like 1.79 (e = 3.2), and the delay will be on the order of 0.2 ns/in. Obviously, it is necessary that the driver circuit be designed like a microwave circuit in the GHz region.

One of the techniques for control of the unit is shown in Fig. 16-3. A temperature-sensing device can be placed on the heat sink of the ILD and the voltage adjusted according to the temperature. Figure 16-4 shows such a temperature-sensing circuit, with a particular arrangement for single-supply operation. With this circuit, one of the JFETs establishes a fixed current through the diode, which acts as the temperature-sensing element. As the forward drop across the diode varies with temperature, it is sensed by the summing gate of U1A whose output swings accordingly. U1B serves to establish an above-ground reference circuit with a constant voltage output. Q1 is connected as a constant current source which renders the current in the diode constant at about 0.21 mA, essentially independent of the supply voltage. However, Q1 current is some-

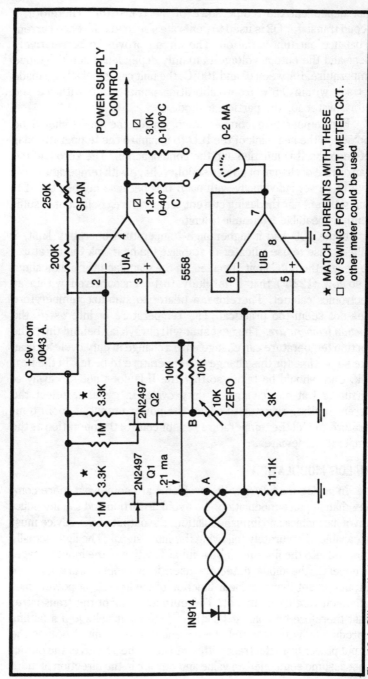

Fig. 16-4. Silicon diode thermometer.

287

what dependent the temperature of the transistor. Therefore, a second transistor Q2 is used to generate a second reference current to balance out this variation. The circuit proves to be relatively linear and the output voltage is mainly dependent upon the diode temperature. Between 0° and 100°C, the unit can generally be made to track within 0.5°C from calibration points made with ice and boiling water for the particular diode.

For temperature compensation usage, the diode should be epoxied to the heat sink of the ILD to minimize the temperature lag between the ILD junction and the control action. The zero and the span or rate of change of power-supply voltage with temperature will have to be set individually with each ILD because neither the ILD lasing current nor the lasing current temperature coefficient is sufficiently repeatable from unit to unit.

If the ILD had a higher high-temperature operating limit, it makes sense to use a heater to stabilize the heat sink temperature. However, the 50°C limit specified on the manufacturer's data sheet is so low (122°F) that it is likely to be encountered within an electronics cabinet. Therefore, a heater to stabilize temperature does not seem too practical. The temperature of interest is the *junction temperature*. The heat sink will always lag behind this. The junction temperature can change with a change in duty cycle and can take some time for this change in temperature to be felt in the heat sink. Care should be taken so that the ILD does not run away at startup and at any change in duty cycle. In general, select the current-limiting resistor to adjust the power output at room temperature and let the temperature sensor control the operation as the temperature deviates.

ANALOG MODULATION

In analog modulation, the problems are somewhat more complex than in pulse modulation. To avoid distortion and the introduction of harmonics and intermodulation, the output of the device must be capable of accurately following the input signal. The light actually launched into the fiber must be a linear function of the input voltage. In general, the diode detectors used in fiberoptic work have an output current that is a linear function of the input light power, plus some constant dark current. The output power of the transmitter must therefore be a linear function of the input voltage if a faithful reproduction is to result at the far end of the circuit. Because the output power from the transmitter can never be negative, the output must assume some median value and deviate in the direction of more and less input power with the application of positive and negative input signals.

Figure 16-5 shows one of the simplest types of drive circuits possible. The curves are for a GE type D42R power tab transistor. This unit is a high-voltage video type used for such things as oscilloscope deflection and video amplification. It actually has a much higher voltage rating than is required in this application, but it is one of the few inexpensive, fast power transistors. The unit has an F_t of 55 MHz and will easily handle an ampere of current. The base current gradations are smoothly distributed and linear in this region on the curves.

Modulation current i_m is the difference between the base current supplied by i_f and the base current required to drive the diode to a given current level as determined from the graph of Fig. 16-5. Note that the load resistor was selected so that at the 500 mA quiescent point, the voltage across the load resistor and the voltage across the transistor are equal.

The plot of Fig. 16-6 shows the calculated relationship between i_m and the collector current. The solid curve neglects the fact that the i_f current flows through the load resistor as well, and the dotted curve shows the effect of this current.

The sharp hook in the data above 800 mA of diode current is due to the fact that the diode and the resistor are using up nearly all of the supply voltage. The curvature of the dotted section results from the base current becoming a significant part of the load current. Between 300 mA and 800 mA, the curve is quite linear. If the diode has a linear relationship between light power output and diode current, the system will function as a very low distortion transmitter/modulator combination. The use of a separate transistor to relay the voltage drop can avoid the curvature in the current curve below 300 mA or at least extend the linear region considerably, since only a fraction of the bias current flows through the load circuit.

One of the first distinctions to be noted between the pulse modulation schemes and analog modulation is that the transistor must dissipate a considerable amount of power. In the example, the transistor is dissipating approximately 3.4W in the quiescent condition. Because of the curvature of the load line, the dissipation drops off to either side of the quiescent point a little less slowly than for an ordinary resistive load. The D42R is rated for 9.6W dissipation with the tab at 75°C. Therefore, the design is reasonably safe. However, a good heat sink will be required for the transistor. It is relatively important that the heat sink for the transistor and the load resistor be

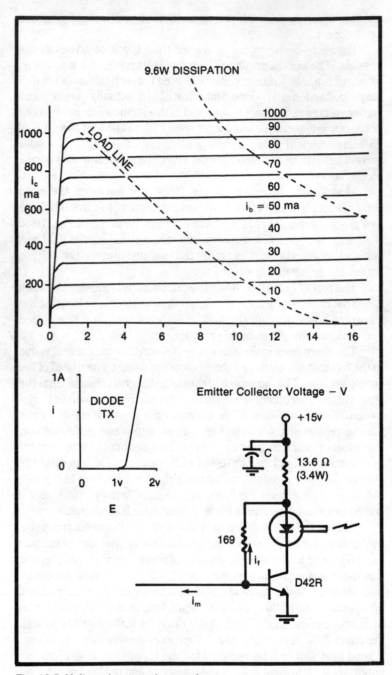

Fig. 16-5. Voltage-to-current converter.

kept separate from the heat sink for the diode because the devices can run much hotter than most diodes can. There is no point in having the drive circuitry heat the diode. Another note of caution is the fact that the resistors used in this circuit—and the pulse drivers, for that matter—must be noninductive if high-speed operation is required. Carbon or metal film devices will work, but a noninductive wirewound will definitely *not* work at frequencies of 1 MHz and above.

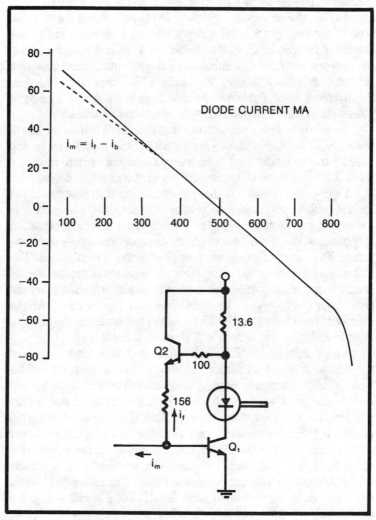

Fig. 16-6. Another voltage-to-current converter.

MODULATION

Several points are to be made on the subject of modulation. One rests with the consideration of the process. As noted earlier, the signal emerging from the diode is best considered as a noise type of signal since it has a bandwidth which is many orders of magnitude broader than the highest frequency with which it can be modulated. The function of the modulation is therefore simply to take this noise band and run its power up and down in accordance with the modulation signal. It generally makes no difference whether a positive-going signal takes the power up or down. However, the signal-to-noise ratio of the receiver falls off sharply when the output is in the low condition. In video applications, therefore, it can make a difference in the appearance of the screen whether the low-modulation condition corresponds to *black* or *white*. In standard TV broadcasting practice, the *reference black* corresponds to a high-level signal. The sync pulses are higher still and therefore are *blacker than black*.

In an analog fiberoptic system that is to have a good signal-to-noise (S/N) ratio, you should try to not let the modulation be too "deep"; that is, let the modulation carry the signal down to too low a level. This invites noise contamination of the negative peaks.

Figure 16-7 shows a technique for testing the response of the system with relatively simple test equipment. The transmitter is modulated with an actual signal or a noise signal, and the modulation is applied to one of the sets of plates of an oscilloscope—the horizontal. The output of a receiver is applied to the vertical plates. The transmitter is first turned completely off to establish the black level and then turned on to maximum power to establish maximum output level from the receiver. The modulation is then applied. At the quiescent point the display will be a dot in the center of the screen. With modulation, the screen will show a diagonal trace.

The illustrations in Fig. 16-7 show the traces for a variety of conditions. If the signal saturates, the top end of the trace will be bent. If the modulation is clipping, the trace will be bent at the bottom end. Nonlinearity will bend the line. Nonlinearity and a phase shift will broaden the trace from a single line. The curve showing 60 percent is probably a desirable case because the transmitter is never shut off and never quite hits maximum power. If the modulation video is closely controlled for maximum amplitude, it is usually permissable to approach the saturating line a little closer and leave a wider margin at the bottom or *black* level. Unlike broadcasting, the overmodulation does not bother other people by splattering the signal into adjacent channels. However the presence of clipping or nonlinearity will load the video with all sorts of products that were not present in the original signal. In particular, if more than one signal is present on the video, the separation between the signals will not be maintained and there will be *crosstalk*.

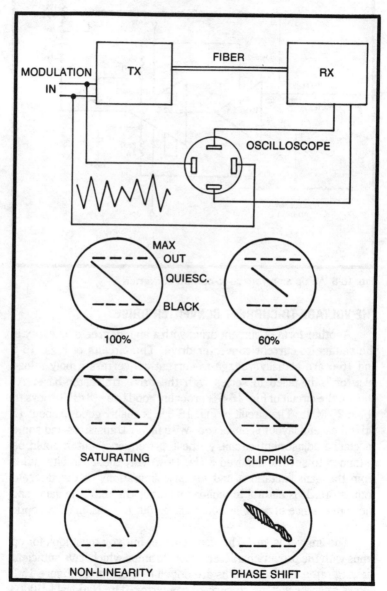

Fig. 16-7. Modulation linearity checking.

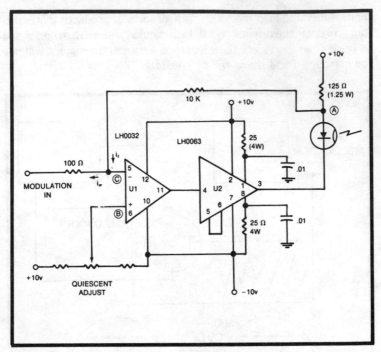

Fig. 16-8. An op amp voltage-to-current converter.

THE VOLTAGE-TO-CURRENT CONVERTER DRIVE

Another form of current drive with a high degree of linearity is the voltage-to-current converter drive. The circuits of Figs. 16-5 and 16-6 are actually voltage-to-current-converters if only a load resistor is installed in series with the drive transistor base. As shown, the circuit of Fig. 16-5 is probably good to a −3 dB flatness to about 20 MHz. The circuit of Fig. 16-6 is probably good to about 12 MHz if careful layout is observed. With faster transistors and some reactance adjustment circuitry, the top frequency cutoff could be extended to several hundred MHz. However, these circuits suffer from the gain limitations and the size limitations of the discrete transistors. It is natural to wonder whether the extreme gains and the convenience of *op amps* could not be put to work in a fiberoptic driver.

The answer is yes! This can be done, but you must look for op amps with the proper characteristics. Op amps which have sufficient slewing rates and bandwidths are not all that common. Figure 16-8 shows a classic voltage-to-current converter. The National LH0063

294

is listed in the catalog as an extremely fast buffer amplifier. This unit is a noninverting current driver amplifier with a current gain of 120 dB and a 6000 V/μS slew rate. It has a power bandwidth of DC to 100 MHz. It will deliver ± 250 mA into a load. With the ± 10-Volt supply, it can provide an output swing of ± 8 volts. The National LH0032 is listed as an ultrafast FET op amp that has a slew rate of 500V/us and a bandwidth of 70 MHz. It requires no compensation for gains above 50. These are some of the fastest op amps available.

In operation, the tendency of U1 is to make the terminals B and C arrive at the same potential. Assume that the quiescent adjustment is set to place terminal B at ground potential when the power is applied. Initially, point A will be at +10V. This will cause a current to flow into the inverting input of U1 and cause the output to slew to a low voltage. The output of U2 will follow this by drawing current through the LED. As this current rises to 80 mA, point A will fall to ground potential and the current through the 10K resistor to point C will cease. The unit is then in equilibrium. Further suppose that the modulation signal applies a positive signal of +0.01V. To drive the input of C to zero again, point A must fall 100 times as far due to the ratio of the resistors in the feedback loop. Therefore, point A must go to −1V, which would require an 11-volt drop across the 125-ohm resistor. This in turn implies that the forward current of the diode would be driven to 88 mA. The device has a current conversion ratio of 0.008A/0.01V, or 0.8A/V.

The nature of U2 is such that with a ±10V supply, the output swing is limited to about ± 8 volts. Suppose that the output pin of U2 rises to +8V and that the forward drop of the diode is 1.2V. Then the drop across the 125-ohm resistor is 0.8V and the diode current would be only 6 mA. To reach this condition, with point A at +9.2V, the 10K resistor is drawing 1 mA, which would require the modulation input to be −0.092V. In the other direction, if the output rose to + 0.06V, point A would have to assume a potential of −6V. Assuming a drop of 2V for the large forward current of the diode, this would place the lower limit of the output of U2 at −8V. In this condition, the current through the resistor would be 10+6/125, or 128 mA. This is probably about all you would like to have the unit draw in any event. However, if you decided that a higher peak power is desirable, this could be accomplished by increasing the voltage of the minus supply. Probably over the range from an input of −0.08 to + 0.06 volts, the current would be a very linear function of the input voltage.

This amplifier, using these units, could probably be made to function from about 0 to 22 MHz with very good linearity. If the diode

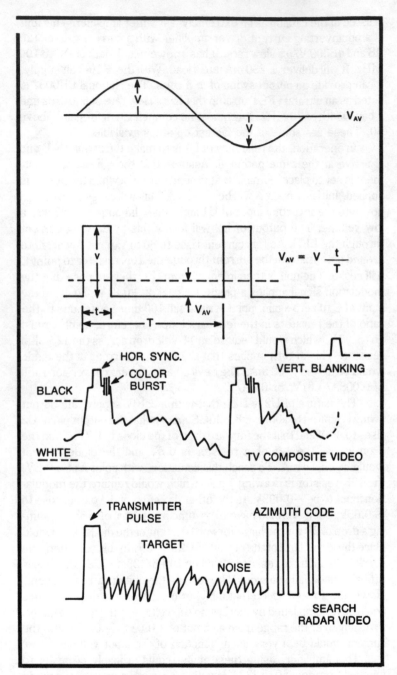

$$V_{AV} = V \frac{t}{T}$$

HOR. SYNC.

COLOR BURST

BLACK

WHITE

VERT. BLANKING

TV COMPOSITE VIDEO

TRANSMITTER PULSE

TARGET

NOISE

AZIMUTH CODE

SEARCH RADAR VIDEO

Fig. 16-9. Average values of video signals.

in use were a Motorola MFOE100, the output would swing between 0.08 mW at an input of −0.092V to 1.4 mW at an input of +0.06V in linear fashion since this diode has a very linear current versus power output curve. As a practical matter it would be better to limit the downward excursion to something more like 0.25 mW, which would be achieved at a swing of − 0.075V on the input.

Note that the swings are not equal, which might or might not prove to be a disadvantage. For example, if the analog signal to be handled by the circuit resembles a sine wave in which the average value is halfway between the peaks, you would want to have the quiescent point centered so that the unit approaches saturation and cutoff at the same time. On the other hand, if the signal were a composite TV video signal that has sync pulses mixed in with the video, the average level is not centered between the positive and negative peaks. Figure 16-9 shows this effect. For the pulse train, the average level of the signal is determined by the duty cycle of the pulses. For the TV video, the average value is determined by both the pulses and the picture content and can have a considerable shift during the vertical blanking period when the signal stays entirely above the black level, except during the NTSC code. For the radar video, the average is also determined partly by the picture content.

For the TV, important information is at both ends of the scale. At the low end, the picture content in a bright picture must not be contaminated by noise. At the high end, the color burst must be preserved. By comparison, the target information in the radar video is mostly at the low level. Therefore, it might make sense to send this signal inverted with the low voltages represented by high power from the fiberoptic transmitter. The transmitter pulse serves only to identify the start of the trace, and the azimuth code is digital and can be clipped without loss of information. In short, there are reasons why the signal may not swing equally about the quiescent level.

Three parameters are available to the designer for the control of the system. The *quiescent current control* shifts the operating current at zero modulation input. The current limiting resistor and the supply voltage also do this but in different directions. Suppose, for example, that you would like to limit the quiescent current to 20 mA but would like to have the peak swing rise up to 200 mA briefly because of the nature of the video. If you leave the positive supply to the system at +10V and take the negative supply to −15V, the output of the unit will be able to swing between +8V and −15V at the diode input. If you assume a 2V forward drop at maximum current for the diode, a 92-ohm resistor will limit the current to 250 mA. At the

20-mA quiescent current, it would be necessary to set point B for a voltage of +0.0816V so that with the input grounded the voltage at C would be −0.0816V below point B. This is an extreme case. The short circuit current-limiting resistor between the minus supply and pin 8 of U2 would have to be changed to 60 ohms. With the circuit as shown in Fig. 16-8, the change in quiescent point could be accomplished by setting the voltage at pin B to +0.075V. However, the maximum current through the diode would be limited to 125 mA if the negative supply remained at −10V.

INTERMODULATION DISTORTION

One of the most common and pernicious defects of an analog system is the occurrence of *intermodulation distortion* (IMD). Intermodulation distortion arises from departures of the system from strict linearity. For example, suppose that you applied two ideal oscillators, A and B, to an ideal resistive attenuating and coupling network. At the output of the network, an ideal spectrum analyzer or selective voltmeter would then find only frequencies A and B and some thermal noise contributed by the fact that the resistors were at a temperature above absolute zero. However, if some of the elements in the network were nonlinear, meaning that they did not respond strictly to Ohm's Law like a diode, etc., you would find a certain number of product terms in the output. In other terms, there would be things in the network which were not inserted by A and B. We would expect to find A+B, A−B, 2A+B, 2A−B, 2B+A, etc.

Figure 16-10 shows the test arrangement. Two oscillators, A and B, are combined through a network and fed to the transmitter over the fiber to the receiver and then into the spectrum analyzer. The output of the spectrum analyzer showing various intermodulation (IM) products is shown below the setup. If the network had been perfectly linear, only frequencies A and B would have been present in the output. The measure of the intermodulation products is taken as the number of dB below the main signals A and B, which are usually equal for this test.

The presence of the isolation pads between the oscillators is a practical necessity. Ordinary oscillators will normally show intermodulation products in the mixed outputs that are no greater than the isolation between the oscillators themselves. The hybrid isolator scheme shown in Fig. 16-10 is an alternative that can give isolations between the oscillators of about 40 dB while inserting a loss of 3 dB.

If the dummy load has exactly the same impedance as the transmitter input, the isolation could theoretically be infinite if the

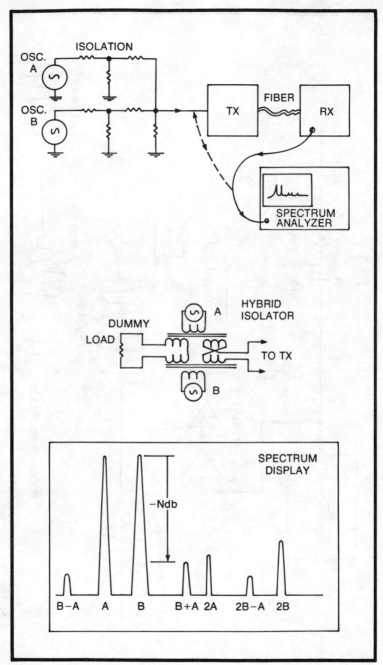

Fig. 16-10. The two-tone IM test.

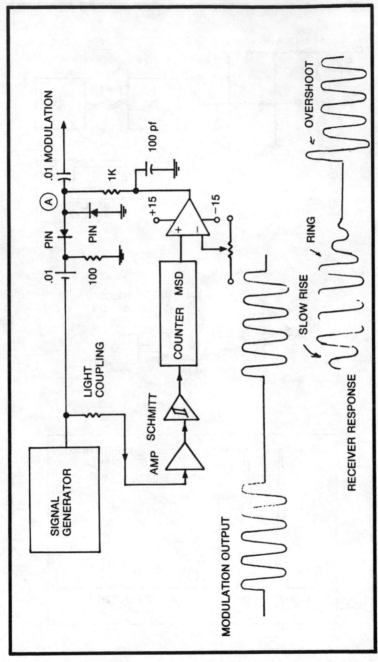

Fig. 16-11. The tone burst test.

two transformers were identical. However, as a practical matter, isolations in excess of 40 dB are difficult to achieve. The isolation can be further improved by using identical attenuator pads between the oscillators and the hybrid transformers. Because the input to the transmitter is probably not purely resistive, it might be necessary to include some reactance in the dummy load as well.

The test shown in Fig. 16-7 is useful principally for aligning the system and setting things up. However, it does not offer much in the way of quantitive information. On the other hand, the IM two-tone test is strictly quantitive and provides numeric output data. It does *not* offer much diagnostic data about why the IM is being generated, though. On modern radio equipment, an IM level of −40 dB is considered to be moderate. Certain equipment is specified to have the highest product no greater than −60 or −80 dB.

THE SINGLE-TONE TESTS

The single-tone test is basically the same as the two-tone test except that only one tone is used. With the spectrum analyzer, the single-tone test will show the harmonics that are generated in the equipment. For military and commercial radios, these are usually specified as no greater than the IM products.

In audio usage, a common specification lumps together all of the extraneous products. This test is termed the *total harmonic distortion test* (THD). In this test, a single tone is applied to the input. The output power is then measured. A very selective filter is next applied to the output, and the input signal is notched out or rejected by the filter and the output again measured. The percentage of THD is equal to the output power with the filter in place, divided by the output power with the filter out times 100. A THD of 0.1 percent is considered to be excellent for a piece of hi-fi gear. Of course, in a fiberoptic link with a bandwidth of 25 MHz, you might have to use a number of frequencies well in the rf range and the filter to notch out the signal must be an rf filter.

The pass band test is accomplished very easily using the spectrum analyzer. The oscillator can be manually tuned across the band, or a sweep frequency oscillator may be used. In the latter case, the response versus frequency curve may be displayed instantaneously on the spectrum analyzer screen. Many of the rf sweepers have an internal detector and require only the services of an oscilloscope.

Another test that is frequently revealing about the operation of a system is the tone burst test. In this test, a short burst of the signal is applied to the input of the system. Figure 16-11 shows a mechanism for generating the tone bursts and the results which may be obtained from an improperly operating system.

A small amount of signal is tapped from the output of the signal generator, amplified if necessary, and squared up with a Schmitt trigger. The counter counts the cycles and when the output goes high, the output of the first op amp comparator goes high. When point A is taken to +14 volts, the series diode conducts and the shunt diode is cut off. In this condition, the diode switch is open and the signal flows through. When point A goes to −14V, the reverse situation obtains. The series switch is biased open and the shunt switch diode is forward biased. About 60 dB of isolation can be obtained in this manner. The arrangement as shown can be made to operate from about a few Hz to 25 or more MHz, depending upon the components in the counter chain. The diodes can be either high-speed switching types or rf PIN diodes sold especially for rf switching.

The receiver outputs shown below the circuit are some of the effects of malfunctions. The slow rise can be the result of a shift in operating point of either the transmitter or the receiver and the overshoot can be caused by the same thing. Overshoot can also be caused by a faulty automatic gain control (agc). The ringing is usually a result of stray circuit parameters such as inductance and capacitance. The tone burst test is particularly informative in a system that handles video. The slow rise can cause blurred or soft leading edges. The overshoot can cause a line on the leading edge of images. And ringing can cause ghosting and trailing edge lines. On a broad-band video system, it is probably a good idea to test with tone bursts at frequencies up to the upper 3 dB frequency. System linearity can also be tested on a point by point basis using the output attenuator of the signal generator and a calibrated meter on the receiver output.

17

Receiver Design

It should be fairly obvious that receiver design is just as important as transmitter design in a fiberoptic system. In the transmitter, the important facets are the problems of obtaining the maximum power output, the desired switching rates for a digital system, and the desired bandwidth free from distortion in an analog system. There are several more requirements in a receiver. The bandwidth or rise time is just as important in a receiver as in a transmitter. Instead of power output, the criterion for good receiver operation is *sensitivity*. The receiver is also charged with distortion in an analog system. In fact, all of the distortions discussed previously required the use of a good receiver. Each of the tests simply lumped together the distortions due to transmitter and receiver. The larger the difference between the power that the transmitter can launch and the power required by the receiver for proper reception, the longer the link can be without repeaters.

There is a significant difference between the economics of fiberoptic systems and radio systems in the fact that the fiberoptic medium over which the message is sent is not free—the fiber cost must be added into the system cost! Often a good transmitter and a good receiver can make it possible to accomplish the overall system goals with relatively cheap fibers.

SIMPLE RECEIVERS

The very simplest of receivers are the phototransistors and the photoDarlingtons. On rather slow systems, a photoDarlington can often be used as the entire receiver because the amplification can be great enough to provide a receiver output directly without any further amplification. The phototransistor operation was described in Fig. 14-8 in some detail and will not be repeated here except to note that a typical phototransistor is principally usable at speeds below 50 kHz. This is more than adequate for a single hi-fi channel of audio or a number of instrumentation applications. It falls far short of video operation, though.

Aside from these limitations, the phototransistor and the photoDarlington have noise and sensitivity considerations that are similar to those of the faster types. We shall be discussing this shortly.

THE PHOTODIODE AMPLIFIER

As noted in chapter 14, the photodiode responds by passing a current in the reverse biased condition which is proportional to the amount of incident light in watts upon the diode. Since the photodiode is essentially a current output device, it usually is operated with a current-to-voltage converter-type amplifier. Figure 17-1 shows such an arrangement. The op amp is operating in a current-to-voltage mode. This is sometimes called a *transimpedance* amplifier.

The operation of the amplifier is such that it tends to drive the difference between the inverting and the noninverting inputs to zero. Therefore, the potential of the inverting input will tend toward ground since the noninverting input is held at ground by connection through R. Suppose that the diode draws a bit of current due to incident photons. This current flowing out of the noninverting input would have to be balanced by a current flowing in through R to make the net current zero at the input. The output therefore has to rise to a level which is sufficiently high to drive a current through R of this magnitude. The inverting terminal will always be at ground potential, and the output will be given by the relationship of Equation 17-1 of Fig. 17-1. This provides the diode with a very small effective load resistance.

The equivalent circuit for a representative high-quality photodiode, the Hewlett-Packard 5087-4204, is shown in Fig. 17-1. It would have been possible to connect the diode in series with a resistor and then take the output as the voltage across the resistor; however, this does not lead to good high-speed performance. For example, at a $-10V$ bias, the capacitance of the diode is 2 pF. To convert the 0.5 uA/uW response of the diode to 5V/uW, you would need a 10 Megohm resistor in series with the diode. Since the shunt resistance is 10^{11} ohms, most of the bias would appear across the diode. However, note that 2-pF in series with the 10^7 ohm load would have a time constant of $2 \times 10^{-12} \times 10^7 \times 2.3$ or 46 μs for 10 percent to 90 percent rise or fall. By comparison, the diode itself has a rise time on the order of 1 ns or less. In a practical installation, the

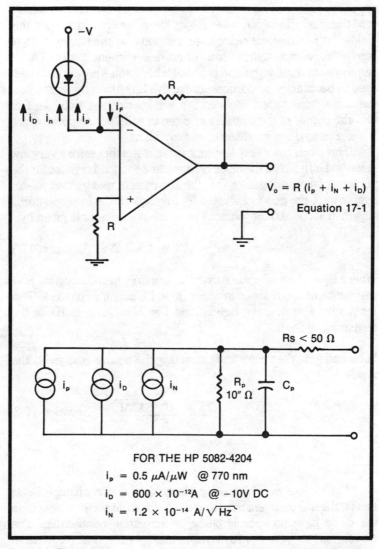

$V_o = R (i_p + i_N + i_D)$

Equation 17-1

$R_s < 50 \ \Omega$

FOR THE HP 5082-4204

$i_p = 0.5 \ \mu A/\mu W \quad @ \ 770 \ nm$

$i_D = 600 \times 10^{-12} A \quad @ \ -10V \ DC$

$i_N = 1.2 \times 10^{-14} \ A/\sqrt{Hz}$

Fig. 17-1. The voltage-to-current converter or transimpedance amplifier.

slowdown would have been even greater since the capacitance of the amplifier used to detect the voltage would be in parallel with the capacitance of the diode.

The response of the diode is in peculiar units, namely amperes per watt of light input. Watts are actually proportional to amperes squared and there is thus a square root relationship involved. In the system, this leads to no difficulty since the output of LEDs and ILDs are more or less linear in watts per ampere. Thus, the two effects tend to cancel. However, when things like noise are considered, this yields the fact that the noise is proportional to the square root of bandwidth when stated in terms of noise equivalent power. This is somewhat at odds with radio practice where noise power is considered to be directly proportional to bandwidth. Along this same vein the name "transimpedance" for the amplifier comes from the fact that the output of the amplifier is given in volts per ampere, which has a dimension of impedance, namely ohms.

This diode has a very low dark current and therefore a very low noise level. In fact, if the diode is connected to a load resistor for the purpose of measuring the noise, the noise generated by the resistor overpowers the noise generated by the diode if the load resistor is less than about 100 megohms. The noise in a resistor is given by:

$$\text{Noise Power} = i_{nr}^2 R = 4 \, k \, T \, \Delta f \quad \text{Equation 17-1}$$

Where i_{nr} = thermal noise current in amperes in the resistor, R = resistance of the resistor in ohms, k = Boltzman's constant, T = temperature in degrees Kelvin, and f = bandwidth in Hz of the measuring system.

Substituting in the values and extracting the square root yields the result:

$$i_{nr} = \frac{1.28 \times 10^{-10} \sqrt{\Delta f}}{\sqrt{R}} \quad \text{Equation 17-2}$$

$$\text{For } T = 25°C = 298°K$$

The various noises or noise sources are entirely independent; that is, the noise current from the resistor and the noise current from the diode have no specific phase or amplitude relationship. The current given by the above relationship is an rms (root mean squared) current with all frequencies within the pass band of the receiver or instrument. The instantaneous amplitude of the current is a continuous stochastic process; that is, all values are possible, but certain values are more probable. Consequently, two noise sources add only in the amount of power that they contribute to the system, not directly in the current. The addition for currents or voltages is given by the square root of the sum of the squares. Thus two 2-ampere noise currents would give the square root of 8 amperes, or 2.83 amperes, rather than 4 amperes, which would be obtained if the currents were of the same frequency and phase.

The noise currents produce the same effect upon the amplifier as the incident light, which gave 0.5 A/W of incident light. Therefore, it is common to refer to the *noise equivalent power* (NEP) of the diode. In complete darkness, the diode still behaves as if it had a certain minute amount of power incident upon it. It is this power which determines the amount of signal that represents the *minimum detectable power* in the system. The 0.5 A/W figure refers to the fact that 81 percent of the photons falling on the active area of the diode generate a hole-electron pair which makes its way to the outside circuit. The remainder are lost in recombinations within the junction.

If you assume that the diode leakage current is 600×10^{-12}A, you can calculate the NEP for the diode. The noise current for the diode is *shot noise,* which is due to the granularity of electrical current due to the charge of a single electron. The concept of an electrical current less than one electron per second becomes meaningless if the period under consideration is a second. Figure 17-2 shows the result of this calculation of NEP as a function of resistance of the load. The statement of the noise from the load resistor overpowering the diode noise for low values of load resistance may be seen from the curve.

At first glance, it would seem that the best sensitivity would be obtained from the diode by using the largest load resistor possible. While this is true, there is another factor to be considered. The internal resistance of the diode is small, but there is also a small capacitance shown in the equivalent circuit of Fig. 17-1. The value of this capacitance is a function of the diode bias. For the 5082-4204, it is 4.5 pF at zero volts reverse bias. It falls to 2 pF at -10V and asymptotically becomes tangent to 1.5 pF, most of which is in the package hardware, at voltages below -20V bias. The problem is that the time constant of the 1.5 pF capacitor in series with the large load resistor is very slow. For example, if you take the -3 dB bandwidth as being the reciprocal of RC seconds, a 10^{10} Ohm resistor would give a time constant of 0.15 seconds for a bandwidth of

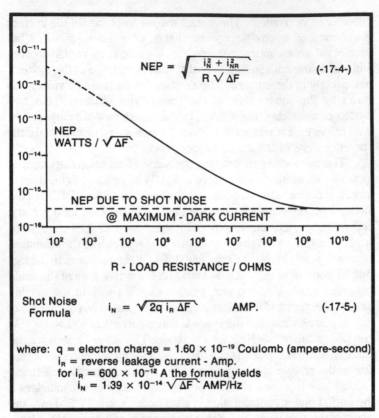

$$NEP = \sqrt{\frac{i_N^2 + i_{NR}^2}{R \sqrt{\Delta F}}} \qquad (-17\text{-}4\text{-})$$

NEP
WATTS / $\sqrt{\Delta F}$

NEP DUE TO SHOT NOISE
@ MAXIMUM - DARK CURRENT

R - LOAD RESISTANCE / OHMS

Shot Noise
Formula $\qquad i_N = \sqrt{2q\, i_R\, \Delta F} \qquad$ AMP. $\qquad (-17\text{-}5\text{-})$

where: q = electron charge = 1.60×10^{-19} Coulomb (ampere-second)
$\quad i_R$ = reverse leakage current - Amp.
$\quad$ for $i_R = 600 \times 10^{-12}$ A the formula yields
$\qquad i_N = 1.39 \times 10^{-14} \sqrt{\Delta F}$ AMP/Hz

Fig. 17-2. Noise equivalent power and shot noise with t equal to 25° C.

6.67 Hz or a rise time of 0.35 seconds. On the other hand, a 1000-ohm load resistor would give a time constant of 1.5×10^{-9} seconds or a bandwidth of 667 MHz and a rise time of 3.45 ns. Thus, the transimpedance gain of the amplifier in Fig. 17-1 is limited by the bandwidth requirement.

The capacitor of the diode equivalent circuit must be charged through the resistance of the feedback resistor of the amplifier. A very small amount of capacitance across this resistor would actually act to flatten the response curve; however, the amount is related to the resistance of the resistor and too much capacitance will actually act to slow the circuit. In practice, the stray capacitance of the resistor will almost certainly act to slow the circuit. For the diode of the example, the rise time is quoted as 1 ns. If the op amp used were an LH0032, the rise time quoted for this device is 8 ns after a delay of

10 ns. Thus, the op amp delay would dominate the situation and you could expect a −3 dB bandwidth on the order of 20 MHz, with appropriate compensation and R equal to 1K ohm. The PIN diode is not the limiting factor in either the noise level or the bandwidth of the system.

THE DIFFERENTIAL VIDEO AMPLIFIER

A number of packaged differential video amplifiers on the market are designed particularly for wide swings, low noise, and good temperature compensation. Many of these are designed to operate without temperature-compensating components or frequency-compensating components external to the package. In general, they are completely symmetrical and are equipped with a gain selection mechanism that operates by selectively shorting the internal emitter resistors. Figure 17-3 shows such an arrangement. These amplifiers generally operate at a low impedance on the input and a somewhat higher output impedance. If the amplifier in the example were a Texas Instruments SN52733 or SN72733 (also sold as a 733 by others), with the gain selected as shown, the output bandwidth would be on the order of 90 MHz. The device has an output swing of something less than 5 volts peak-to-peak or about 1.2V rms. This translates to 0.012 volts at the input, or 240 μA rms. At 0.5 A/W, this would correspond to a diode input power of 490 μW optical. This

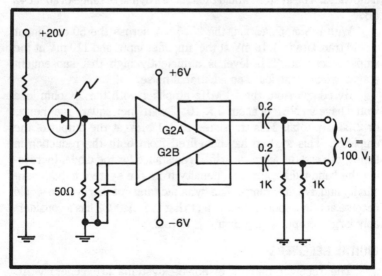

Fig. 17-3. The differential video amplifier.

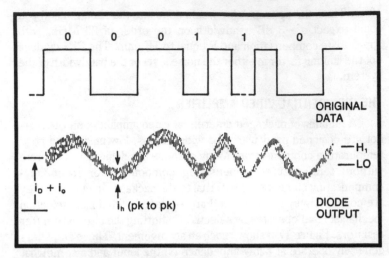

```
    0     0     0  1     1  1     0
```

ORIGINAL
DATA

—H₁
—LO

$i_D + i_o$

i_h (pk to pk)

DIODE
OUTPUT

Fig. 17-4. Digital data degradation. The diode output is degraded by attenuation and dispersion in the fiber itself and is loaded with shot and resistor noise. The leakage current offsets the center line of the data.

level would represent something like the saturation level optical input. From the chart of Fig. 17-2, you can see that the noise power would be something like 3×10^{-12} watts per square root hertz. Translated over a 90-MHz bandwidth, this is an NEP of 28.5 nanowatts. The system would thus have a dynamic range of about 42 dB.

With the amplification, the 28.5 nW across the 50 ohm input would translate to 1.19 mV at the amplifier input and 119 mV at the amplifier output. This level is sufficiently high that subsequent stages would contribute no additional noise.

By comparison, the 20-MHz amplifier with the 1K ohm load would have an NEP of about 5×10^{-13} watts per square root hertz, or 2.23 nW total. This translates to 1.12 μV at the output of the amplifier. This system has benefited from both the reduction in bandwidth and the increase in load resistance for the diode. Increasing the bandwidth extracts a penalty from the system in both sensitivity and dynamic range. The dynamic range of this system would receive another boost from the fact that the LH0032 has a considerably larger output swing than the 72733.

DIGITAL RECEIVERS

The receiver outputs of analog systems are generally AC coupled, and the level restoration is usually accomplished at relatively high levels by clamping. However, in the digital system it is usually necessary that the pulse train be recovered with regard to some absolute level. This can play into the detection problem in several ways.

Figure 17-4 illustrates this phenomenon. The original data is presumed to have emerged from the transmitter in fairly square form in the NRZ/FM format shown in Chapter 3. After some transit through the fiber, the corners have been rounded on the received wave and the noise has been added by the diode output. In addition, the diode and resistor have added noise and leakage current. In order to accurately recover the data, there must be *slicing levels*, designated as HI and LO in the figure. When the net signal is above HI, it is considered high. When the signal is below LO, it is considered low. There must be some gap between the two; otherwise, the system noise could output a random stream of highs and lows even with no input to the diode at all.

We discussed earlier that i_D is a function of temperature. Suppose that for the 20-MHz system, the initial leakage current starts at 600×10^{-12}A at 25°C. For this level, the noise current from the 1K ohm resistor is 1.81×10^{-8}Arms, or 5.12×10^{-8}A p-p. If the temperature of the diode junction increases to 50°C, the leakage current will rise to 6.28×10^{-9}A, thus shifting the base line. The resistor noise at the same temperature will rise slightly to 1.89×10^{-8}A. Even worse than either of these is the fact that the input offset current of the amplifier is likely to drift by one-half percent/°C. If this started at 25 nA, it could very easily have climbed to 28 nA at 50°C. This change is almost equal to the resistor noise swing.

HI and LO are to be set levels, and they must be placed several times as far apart as the actual noise swing if any significant change in temperature is to be accommodated. The alternative is to let HI and LO float together about the signal level. The transmitter diode output has a tendency to decrease with time, so this will also affect the signal amplitude.

Figure 17-5 shows one technique for providing a tracking control to eliminate the effects of drift. Amplifier U1 acts in the normal transconductance function. Amplifier U2 acts as a low-pass filter establishing an output voltage which is locked to the average output level of U1. U3 and U4 can be relatively slow op amps that sum and invert an offset from the average level of U1 to supply a reference voltage for the comparators U5 and U6. Point A will be a little below the average output of U1 and point B will be a little above. When the output of U1 is slightly above point B, the output of U5 will go low and

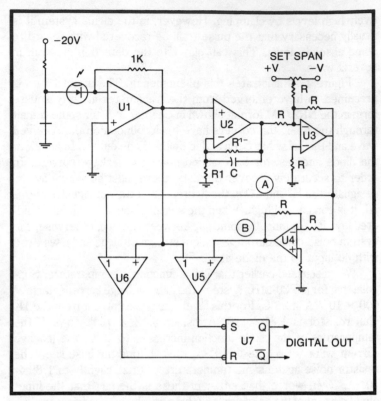

Fig. 17-5. A drift-compensating digital receiver.

set the flip-flop. When the output of U1 is slightly below point A, U6 will go low and reset the flip-flop. The width of the deadband between HI and LO is set by the amount of positive voltage applied by the *set span* resistor. If the offsets of U2, U3, and U4 are small, the deadband will accurately track the average output value of U1, and the deadband will be symmetrically arrayed about the average. This amplifier arrangement is not very well suited to plain NRZ modulation of the system since a very long string of zeros or ones could cause the output of U2 to drift toward the zero or one level. For true NRZ operation, you would have to supply a fixed reference for the comparators and depend upon a somewhat wider swing from U1, which implies a larger operation threshold power or less sensitivity.

One of the most significant differences between the fiberoptic system and the typical radio system is the fact that the fiberoptic system is completely free of outside noise and the system is in general entirely static. In a mobile radio link, for example, the radios in the cars may receive input signals measured in volts or tenths of volts when the car is near the base station and may receive signals of a few tenths of a microvolt when the car is in a remote position or is shadowed by hills, buildings, etc. Furthermore, when the car is in motion, the signal may fluctuate between very strong and very weak levels with only short spacing, especially if the car is traveling through a multipath interference pattern. The fiberoptic system is usually quite static, though. The path loss is generally fixed by the various coupling and attenuation losses built into the system, and these change very slowly with age, if at all.

Similar considerations apply to noise. In the mobile radio system or a fixed microwave link, for that matter, the system can be exposed to extraneous noise generated by appliances, poorly shielded automotive ignition systems, corona from power lines, etc. In distinction, the noise within a fiberoptic link is due entirely to the components of the system itself. There is essentially no pickup of outside noise whatever. The noise can have some sensitivity due to temperature and so, but this will generally follow a simple repeatable pattern.

The net result of these factors is that fiberoptic systems can be designed in a considerably simpler fashion than typical radio links for any one application. A manual adjustment of sensitivity will probably be quite adequate for a given installation. Only when the transmitter and receiver are to be supplied as general products in a wide variety of systems is there any great advantage in having wide-range self-adjusting features. The commercially available transmitters and receivers are often equipped with an agc system, simply because they become more flexible and easy to use. The purchaser may expect to buy the units and repeaters and cable to install a system. If the system is longer than can be handled without repeaters, the economic considerations will dictate that he buy as few repeaters as possible or the lowest grade cable which will handle the signals. Thus, it is not unlikely that he will attempt to use the receiver as close to the noise level as feasible. The receiver that can work close to the noise with minimum data corruption will therefore receive a certain amount of preference, provided that the increased cost does not negate the advantage.

THE AVALANCHE PHOTODIODE

The avalanche photodiode actually functions in much the same manner as the PIN photodiode, with the exception that it requires a

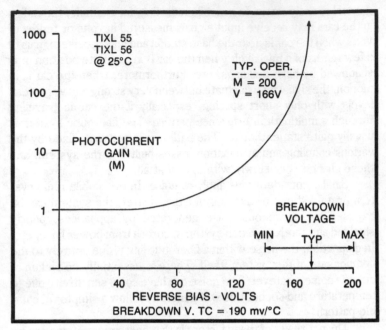

Fig. 17-6. The APD operating curve.

much higher bias. The use of transimpedance amplifiers or load resistors and voltage mode amplifiers is about the same. Thus, the circuitry employed is similar. However, there are differences which must be accounted for.

For one thing, the APD is much more sensitive than the PIN diode. Whereas the responsiveness of the PIN is typically on the order of 0.5 A/W, the responsiveness of the APD can be adjusted by tailoring the reverse bias to range between 5 and 100 A/W. As a matter of fact, there is always the possibility that the too high a bias can cause the APD responsiveness to go to infinity. Of course, this is not desirable because it means that the first photon to hit the diode sends it into a *stuck low* condition.

This situation is compounded by the temperature sensitivity of the APD and the very steep operating curve. Figure 17-6 shows the "typical" curve for a pair of Texas Instruments silicon photodiodes. The TIXL 55 is mounted in a pill package for microwave stripline, and the TIXL 56 is mounted in a metal TO-18 package. Both have lenses. The photocurrent gain, M, is very steep in the vicinity of the operating point. Between the 166V operating point and the 170 V breakdown point, the dark current in the diode, i_R, climbs from 10 nA to 10 μA, three orders of magnitude. Even worse than this is the fact that the 170V number is very nominal indeed, because the product is only specified to have the breakdown voltage somewhere between 140 and 200 volts. Furthermore, the breakdown has a nominal temperature coefficient of 190 mV/°C. If the temperature rises from −20°C to + 60°C, the breakdown voltage will swing by + 15.2 volts! It is fairly obvious that the avalanche name is appropriate. The normal operation of these units finds them teetering on the edge of a cliff.

The avalanche diode is best operated in some kind of a constant current mode. However, the dark current alone cannot necessarily be measured when the diode is operating with incoming light signals, particularly in an analog system where the average level of the photocurrent will be a function of the video content.

To further amplify this point, assume that you operated the diode with a constant forward current of 5 μA and that the typical curve applies. Then, with no light input, the unit would be operating at something in excess of 169V, and the photocurrent gain would be on the order of 800. On the other hand, when enough light input arrived to produce the entire 5 μA, the forward drop of the diode would be down around 160V and the photocurrent gain would have fallen to something like 20. This is a powerful nonlinearity or agc action within the diode itself. Because it could operate at the full speed of the diode, it would distort any analog data very severely. On the other hand, in a pulse system the nonlinear distortion would not be very severe and the effect of the agc might be quite attractive in some circumstances. Of course, if the variation in the supply were slowed so that it could not follow the video or if the percentage of modulation of the analog video could be reduced, the distortion could also be reduced.

The characteristic of the avalanche photodiode is such at the lower gains, the signal-to-noise ratio varies rather slowly, decreasing slightly with increasing gain. There is then a rather abrupt departure at some gain M_t, beyond which the gain begins to become more granular or noisy and the signal-to-noise ratio rapidly deteriorates. For the diodes in this example, M_t is measured with V = 100V and 1 nA of photocurrent. In this condition, the value of M_t is quoted as 100 minimum and 200 typical. This does not seem to jibe with the voltage versus gain curve, which seems to indicate that the gain is something more like 2 at 100V reverse bias. However, the variations in gain with voltage are very rapid and the presence of the photocurrent will tend to knock down the forward drop considerably. For the best signal-to-noise ratio, and in general for gains less than M_t, the NEP of the diode is quoted at 10^{-12} W/square root Hz. It is common to assume that the APD will have a 15 to 20 dB S/N advantage over the PIN. This advantage is not borne out by a comparison of the figures for the example diodes and the PIN system with 20-MHz bandwidth quoted earlier. The reason for the disparity is that the APD does not really begin to show the large advantages until very wide bandwidths are reached. The diodes in the example have a gain bandwidth product of 80 GHz, and the quoted figure for NEP is taken at 1 GHZ. In order to do the same thing with the PIN, you would need to have a load resistance of less than 50 ohms. The advantage would indeed be on the order of 15 dB for the APD. Of course, the PIN cannot reach 1 GHz due to rise time restrictions. Therefore, the comparison is questionable. For the APD, the signal current roughly increases as M. The noise increases roughly as $M^{1.52}$ up to the M_t point, after which it deteriorates rapidly.

Texas Instruments supplies a full detector head unit, the TIXL73, which is equipped with a 50-MHz transimpedance amplifier and a regulated avalanche bias gain control circuit. The unit is equipped to operate with the TIXL 70, 71, and 72 packaged detectors. These detectors not only contain the APD but they also have a built-in reference diode whose temperature coefficients are matched to the APD. They also contain a Peltier cooler to permit operation at low temperatures for the APD. This is the easiest way out of the problem of trying to make use of the very fast response and the enhanced sensitivity of the APD. The design of a suitable control for APD usage in general case situations is simply not child's play. The TI units are intended primarily for laser range finders and optical communications units where a large range of signal levels and characteristics must be accommodated. Fortunately, as noted before, the typical fiberoptic system has only static characteristics and can thus be accommodated by something less sophisticated.

Figure 17-7 shows one technique for measuring stabilization and compensation. It requires an extra APD. In the circuit, two APDs, D1 and D2, are biased from a −300V source through a pair of 300V transistors. One of the APDs is used as the signal detector. The other is used as a bias reference. The two diodes should be mounted on the same heat sink and should be as alike as possible. A LED labeled D3 supplies the reference APD with a photocurrent equal to the photocurrent in the detector APD. The regulation amplifier is connected in a common-mode rejection fashion and operates at a gain of 0.1. In effect this unit will deviate its output voltage from the voltage setting of the tap upon the current adjust tap by a voltage equal to one-tenth of the potential difference between A and B, which is a measure of the current through the reference APD.

Suppose you wish to operate the APD detector at an average current of 500 μA. In this case, substitute a resistor for each of the APDs. Suppose that the APD will draw 500 μA with the photocurrent at a voltage drop of 170V. In this case, the equivalent resistance of the APD is 340K ohm. With a 340K ohm resistor in place of each

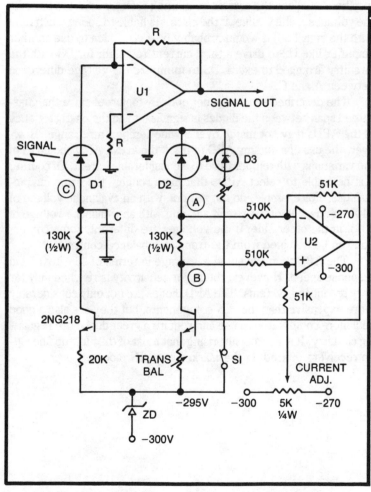

Fig. 17-7. An APD bias control technique.

diode you can adjust the current adjust setting and the transistor balance so that the two 130K ohm resistors each show a voltage drop of 65 volts, indicating a current of 500 μA in the leg. The process is slightly iterative; that is, the transistor adjust is tweaked and the current adjust must be reset to arrive at the final setting for each. The APDs can then be replaced. If both diodes are dark, the voltage corresponding to A will be pushed to a relatively high negative value as the diodes attempt to draw the 500 μA. However, this will only push them into a more noisy avalanche region and they will draw only the 500 μA. With the light signal on, the circuit to D3 may be closed and the current regulating resistor adjusted to make the voltages A and C alike. If the circuit is liable to spend much time with the signal off, it would probably be a good idea to use another amplifier like U2 to drive a third current regulator for D3 with the circuitry arranged to excite D3 to minimize the voltage difference between A and C.

The described scheme is not entirely foolproof since the guaranteed span between the diodes is significant and the characteristics of the APDs may not match over a wide temperature range. However, the use of a dummy APD to track the active APD should calm the variations with temperature to a considerable extent. Of course, it is desirable to select APDs that have roughly the same characteristics. For example, do not put one with an avalanche voltage of 140V in one side of the circuit and one with an avalanche voltage of 200V in the other side. If the voltages are different, some correction can be obtained from the transistor balance control.

The APD then has some advantage in terms of sensitivity and response speed. However, the major sensitivity is reached only for very broadband systems. The APD diodes are not only considerably more expensive than the PIN components, but they are also a good deal more complicated to use and require a great deal more regulating circuitry. It is not too surprising that most of the built-up fiberoptic receivers offered on the market use PIN diodes.

18

Noise Degradation

To the communications engineer, anything which is present in a signal that interferes with the proper and flawless detection and replication of the original data is *noise*. This may be something as slow moving as the base line drift of an analog voltage signal or as fast moving as the 4-MHz *snow* that appears on a weak TV signal. It originated with the audible noise on telephone circuits; however, today the definition has been expanded to describe any random interference.

The definition of noise as an unwanted signal occasionally runs into problems, at least at the philosophical level, because sometimes the "noise" is not unwanted. For example, the incandescent lamp which illuminates this paper as you read it has a "noise output" in that the photons emitted from the filament boil off at random intervals and essentially random energy with only the statistical distribution determined by the temperature of the filament. However, this illumination is certainly not unwanted! Furthermore, the output of an LED or an ILD is noisy in that the radiation is not a single fine spectral line, but is rather a band of radiation. The radiation is just as incoherent within the band as the incandescent lamp, but the statistics of the spectrum are more confined for the LED and very much more confined for the ILD.

In any communications system, some sources of noise contaminate the data and ultimately limit the capabilities of the system. Figure 18-1 depicts this situation. On this graph, the Y axis represents power level in dbW, and the X axis represents an arbitrary

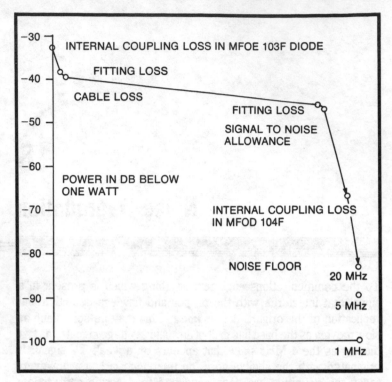

Fig. 18-1. System power budger.

scale of position through the system. The system starts with a Motorola MFOE 103F LED source. This unit is packaged in an Amp connector and can be driven by a circuit like that of Fig. 16-1 or 16-8 for pulse or analog usage, respectively. You can use the MFOD 104F for the receiver. The curve of Fig. 18-1 shows the coupling losses internal to the fitted diodes. Obviously the coupling loss on the receiver end must be added to noise floor because the signal will have to be stronger to overpower the internal noise. The receiver noise floor is shown for several levels of bandwidth but the remainder of the system is shown only for the broadest bandwidth.

The main factors about this figure are the output of the source and the noise in the receiver as translated through the coupling loss. In the example shown, only 6 dB is left for cable loss with a system bandwidth of 20 MHz. Both of the diodes are effectively pigtailed into 200 μm fibers as purchased. Therefore, the joint loss should be small. If the cable used were a Belden 220001 with a loss of 10 dB/km maximum, the cable length could be up to 0.6 km.

If the system were backed down to a less ambitious bandwidth like 5 MHz or 1 MHz, a resulting increase in length for the system would be possible. The noise falls for two reasons. First of all, the load resistance in the transimpedance amplifier can be raised because the response no longer has to be so fast. This reduces the resistor noise. Secondly, the reduced bandwidth also reduces the noise current. Increasing the system bandwidth has a powerful effect upon the cost of the system, particularly if the system is longer than can be handled with a single link and repeaters are required.

A very noticeable item in the budget is the allowance for signal-to-noise ratio. You might question the validity of including 20 dB for this item. Is this arbitrary or realistic? Perhaps the best way to answer this question is to examine some of the signal-to-noise ratios encountered in typical systems. Because the effects of noise are somewhat different, we shall consider the effects of noise degradation separately on various types of digital and analog systems.

ANALOG SIGNAL-TO-NOISE RATIO EFFECTS

The effects of noise degradation in an analog system are very much subject to interpretation. What one person might consider an acceptable degradation may not be acceptable to another. We shall treat a number of examples separately and attempt to explain some of the results of some arbitrary levels of noise degradation.

ANALOG INSTRUMENTATION

In this type of instrumentation, a remote instrument, such as a thermometer, provides an analog output voltage that is translated into a light power and then retranslated into a voltage which is a replica of the input voltage with errors contributed by noise. Because this type of system is seldom readable to better than 1 percent, a 40 dB signal-to-noise ratio is required to provide errors equivalent to other reading errors. For a five percent readable system, a 26 dB signal-to-noise ratio would be required. This type of system is usually very slow. Therefore, a very narrow bandwidth is usually acceptable and the high signal-to-noise ratios are usually easily attainable. Note that such items as thermal baseline drift can easily promote errors larger than one percent to five percent.

HUMAN SPEECH

Human speech is a very subjective thing to interpret. A communication link which one person finds easy to interpret might produce nothing but meaningless noise to another because of var-

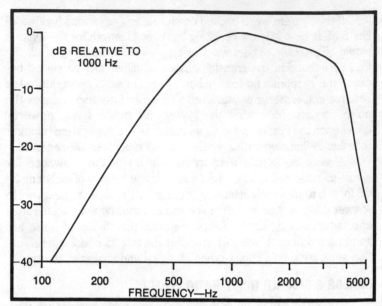

Fig. 18-2. Telephone line standard pass band.

iations in hearing ability and knowledge of the subject matter. It is not unusual to find two people carrying on a conversation at a cocktail party with a signal-to-noise ratio on the order of −20 dB. However, the information being transferred is generally trivial, the people can usually see one another, and each generally has an idea of what the other person is trying to say and so does not have to catch every word.

The Bell System has put a great deal of study into the amount of acceptable noise on telephone lines. The pass band of the typical telephone line is shaped, so a uniform noise source will not contribute the same amount of noise along any given frequency interval. Figure 18-2 shows this standardized pass band. When uniform noise is passed through a filter having this pass characteristic, it is referred to as a *C-notched noise*. This pass characteristic shows the way in which the noise presented to the speaker's ear will be shaped.

When a test is made of the signal-to-noise ratio for a single tone at 1 kHz with a transmitted power of −12 dBm, the received signal-to-noise ratio varies with the connection length as shown in Table 18-1. From Table 18-1, you can see that a signal-to-noise ratio of 30 dB would be considered marginal performance for a long-distance link for voice communications.

Table 18-1. Signal to C-Notched Noise Ratios.

SIGNAL TO C-NOTCHED NOISE RATIOS		
CONNECTION LENGTH	MEAN	STD DEVIATION
SHORT	42.1 db	13.0 db
MEDIUM	36.5 db	5.3 db
LONG	35.4 db	3.8 db

It should be noted that the optimum design of a voice analog link is a very sophisticated subject which is well beyond the scope of this book. The Bell system publishes a series of manuals and standards on the subject which provide a great deal of information. If you intend to build a fiberoptic voice link, refer to these standards.

HI-FI

A quality hi-fi system will generally have the −3 dB points at 20 Hz and 20 kHz and is thus much broader at each end of the spectrum than a telephone network. In addition, there may be weighting or preemphasis applied to the signals. The FCC requirements for FM broadcasting are that the signal-to-noise ratio must be below −60 dB in the band from 50 to 15,000 Hz for a 100 percent swing on a 400 Hz tone. The 100 percent swing means that this is the loudest tone which can be sent through the system without distortion.

The −60 dB below the largest undistorted signal represents fairly good hi-fi performance. If the fiberoptic link can achieve this S/N ratio, the contribution to the system noise will be acceptable in most cases.

TELEVISION

For TV, the standard usually followed in CATV installations comes from the IEEE transactions on broadcasting. This requires that the sum of all noise contribution including co-channel interference, adjacent channel interference, other extraneous signals, and system generated noise shall be at least −40 dB below the peak carrier level in each channel. The white level reference on TV is 12.5 percent ± 2.5 percent of the peak picture carrier. Therefore, a noise level which is −10 dB will be roughly comparable with the white level and the picture looks very noisy. At −20 dB, the noise will be noticeable in certain areas of the picture and the quality is still not very good. At a −30 dB S/N ratio, the noise will generally be noticeable only in non-busy areas of the picture.

The interference from other channels is more noticeable at any

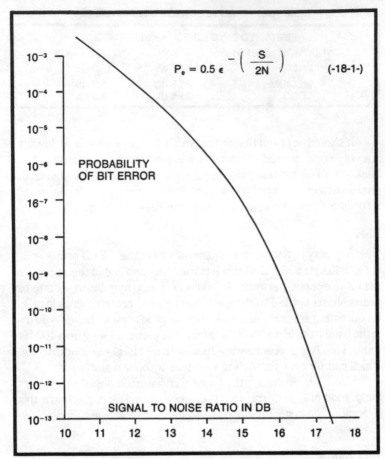

$$P_e = 0.5 \, \epsilon^{-\left(\dfrac{S}{2N}\right)} \qquad (\text{-18-1-})$$

Fig. 18-3. Probability of error in a noncoherent FSK system.

given level because the products tend to be stationary on the TV screen. The noise generated by a fiberoptic receiver is generally diffuse enough to be nonstationary on the TV screen and is therefore less objectionable at a given level. However, even in a system where only one channel is accommodated, a signal-to-noise ratio of 40 dB is required for a really crisp and clean picture.

The limits presented in this brief treatment are intended only to serve as a guideline for system design margins. If your final product is to interface into a telephone or broadcasting network, you must comply with the appropriate specifications in terms of introduced noise and distortion. This means that a detailed study of the applicable specifications is always required.

DIGITAL NOISE EFFECTS

There is no real requirement in a digital system that the receiver accurately reproduce the waveform at the transmitting end. In fact, most digital systems actually clip and reshape the waveform in a clean up operation, as shown in Fig. 17-4. These operations will generally serve to restore the rise times of pulses and flatten both the zero and the one level so that the waveform is fully restored.

At least in theory, a noisy and degraded digital signal can be reprocessed and restored to its original condition in a completely error-free way, because the exact height of a wave carries no information. This is one of the chief attractions of digital data transmission. With an analog link, any noise degradation of a signal simply stays with the signal and cannot be removed. However, the digital signal can be reshaped to restore the original signal with no loss of information.

On the other hand, there is some level of signal degradation where a digital signal has deteriorated to the point where the receiver can no longer distinguish whether it is a zero or a one, or at least the receiver will make an occasional mistake. After all, the presence of noise on the signal can make the low level appear higher than it should be or make the high level appear lower. A noise burst on the leading edge of a pulse can make the pulse appear to be later than it should be or earlier, or it can make the pulse appear either broader or narrower than it was. You might therefore expect that there is a threshold where the digital data rather rapidly goes from an essentially error-free condition to a very unreliable state. This proves to be the case.

P. F. Panter has solved the probability of error for an FSK (frequency-shift keying) system. This system is not identical to the FM/NRZ system described in Chapter 3. However, there are sufficient similarities so that the probabilities of error are similar. The curve of Fig. 18-3 shows this simplified formula and the result. At 10 dB S/N, the probability of error is unacceptably high for nearly any application with an error probable for every few hundred bits. However, by the time that the S/N reaches 16 dB, the number of bits with one probable error has grown to 10^9. By 18 dB, the rate is nearly 10^{15}. The number of bits between probable errors is called the *bit error rate*, which is usually abbreviated BER.

To get a feel for these numbers, suppose that a system transmits data at a rate of 10^8/bits/sec and the system bit error rate is 10^{15}. You would expect the system to make an error on one bit every 10^7 seconds, which is 115.75 days! Without getting really mathematical, it can be shown that a 20 dB signal-to-noise ratio using nearly any practical encoding scheme will give bit error rates that are acceptable for nearly any application. As the S/N ratio falls below about 16 dB, the BER rapidly deteriorates into an unacceptable condition.

As a practical matter, any time that the BER exceeds something like 10^{10}, it is wise not to depend upon the arithmetic too much since the probability for error in some of the remaining system will probably begin to dominate the system. For a well designed receiver, a S/N ratio of 20 dB will generally be more than adequate for nearly any purpose, and an increase in S/N ratio is not likely to provide any practical, measurable improvement in BER.

ANALOG VERSUS DIGITAL MODULATION

Offhand, it would seem that the digital system holds all of the cards. Not only is it possible to restore the data perfectly, but it is possible to do so with a signal which is only modestly above the noise. An analog TV signal needs something like a 40 dB signal-to-noise ratio, and at that it is slightly degraded in a way which can never be restored. By comparison, an 18 dB S/N ratio in a digital system provides fully restorable data which can be passed through thousands of repeaters without any detectable error. With even a dozen −40 dB S/N ratio analog repeaters, the data will be irreparably degraded. Why does one see analog systems in use at all?

The answer to the question lies in the fact that the advantage is not all that simple. Figure 18-4 shows an analog signal that is being converted to a digital signal. This is usually done with an analog-to-digital (A/D) converter. Because it takes some time to send the data, there is always a *sampling time error*. In addition to this, the A/D converter will have only a finite number of bits, or a finite resolution. Therefore, there will always be some quantization error or *quantization noise*.

19

Accessories

One of the areas where the infancy of fiberoptic devices is most evident is the area of accessories. Many of the most common items which permit the rapid assembly of waveguide systems in the microwave portion of the spectrum either have no counterpart or are not commercially available in fiberoptic components. There are a number of reasons for this, but one of the strongest is the lack of standardization.

CONNECTORS AND JUNCTIONS

There is a substantial lack of standardization in fiberoptic cables and connectors. In single-fiber cables, for example, the Hewlett-Packard connectors have a 10-32 thread and a protruding barrel which centers itself in the bored fitting part with the male position of the connector attached to the cable. On the other hand the SMA-type connectors by Spectronics and several others have a larger thread and less protrusion. And none of these will mate with the AMP series, for which the Motorola sources and detectors are designed.

Probably because of this, you cannot simply go out and shop cables from one vendor, connectors and tees from another, and diodes from a third. You must carefully obtain components and hardware that will fit together.

A particular case in point is the tee connector or power splitter. In rf and microwave systems, a single source must be capable of driving several receivers, loads, or instruments. Figure 19-1 shows

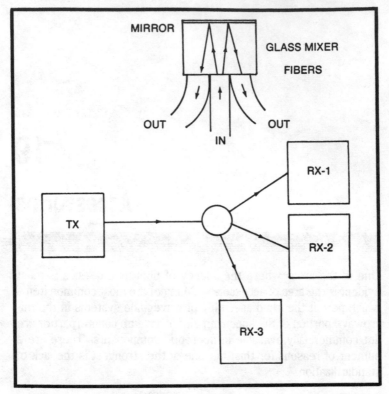

Fig. 19-1. The star connection.

a simple system from an ordinary rf or video standpoint. This arrangement is termed a *star connection* because the signal travels into the center of the star and then spreads radially outward to the various detectors. The detectors could be a data receiver, a power monitor, or a data analyzer. Or they simply might represent three different places where the same data from a single transmitter is required. The former case represents an instrumentation problem, and the latter could represent an operating system problem such as CATV distribution.

In theory the star connection should supply each branch with the same power so that each output receives 1/N of the input power. In fact, the simple arrangement shown in Fig. 19-1 would really not do such a good job of this. With the arrangement as shown, most of the power reflected by the mirror would bounce right back down the center cable, since the points of reflection for the outer cables lie off axis far enough for the reflection point to be in the reduced intensity portion of input fiber pattern. There are perhaps some ways to cure this. For example, if the mirror were conical with the cone pointing toward the axis of the input fiber in the center, the split might be more even.

The type of star connection shown has a more serious problem when you consider the situation in which any of N transmitters have to access to a single receiver. In this case, each of the peripheral transmitters would waste a good deal of its power into the other transmitters, rather than the central receiver fiber. The conical mirror might help this to some extent. Unfortunately, the technology to manufacture such a device with the very tiny dimensions required to match the dimensions of the fibers is not available, at least not on a low-cost commercial basis.

THE LINEAR SYSTEM

The star connection is principally useful in locations where it is desirable to tie together a number of instruments that are physically distributed about a central point. There are a great many situations where this is simply not the case. For example, in a CATV system or a telephone system, the various subscribers are distributed in a more or less linear fashion. They may be distributed for several miles along a road with perhaps 100 yards between the subscribers. Obviously, it is not very economical for fiber to go from the transmitter site at one end of the road to a point halfway down the route and then bring cable back from the central point to each separate subscriber. A far more economical system would simply be to pass the single fiber down the road with each subscriber tapped across the fiber where it passes the subscriber location. This is termed a linear or *daisy chain* installation. Figure 19-2 shows such a connection.

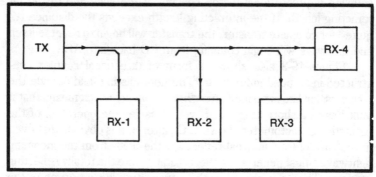

Fig. 19-2. The daisy chain connection.

One of the facets of this system is that the amount of coupling to each subscriber is different. Consider the situation shown in the figure. Subscriber 1 is relatively close to the source; therefore, there has been little attenuation along the line. For subscriber 2, there is more attenuation, and there is also the power which subscriber 1 has removed. Subscriber 4 is in the position that he must not only face all of the attenuation but also must face the power removed by subscribers 1, 2, and 3. If the signal strength at each of the subscriber terminals is to be equalized, each must have a different coupling to the line.

In rf and microwave work such coupling is usually provided by means of a directional coupler. Figure 19-3 shows the action of a directional coupler. The electrical analog is shown for a pair of two-wire transmission lines; however, the principle also works with waveguides and coaxial lines. Line AB in this example is energized with a current. As shown earlier, this current sets up a magnetic field circling about each of the conductors. However, the magnetic field does not immediately terminate in the space outside of the line pair, and a small amount of the field will link conductor pair CD, thereby inducing a field in CD. If the coupling between the two lines is light, a large amount of power will eventually be transferred from AB to CD and the forward wave will travel in the same direction if the velocity of propagation is about the same in both lines.

Your first impression might be that the coupling would build until the current in CD is equal to the reduced current in line AB; however, this is not the case. The wave induced in CD is in quadrature with the wave in AB. Therefore, the wave induced from CD back into AB is in phase opposition to the wave originally in AB. Instead of the process proceeding to equilibrium, it proceeds to completion with all of the power winding up in CD. The amount of the coupling is proportional to the spacing of the lines and to the interacting length. If the interacting length exceeds the distance required for complete transfer, the transfer will begin to reverse itself and the wave will begin coupling from CD back into AB.

Figure 19-3 also shows a form of directional coupler constructed in stepped index fiber. The nonradiating field outside the transmission line is termed an *evanescent mode*, which means that it vanishes in a short distance. If you consider the operation of the extinction refractometer shown in Chapter 8, it is obvious that even in the case of total internal reflection, the fields from the incoming lightwaves must penetrate a short distance into the totally reflecting medium. Thus, if a short length of two fibers has a portion of the

330

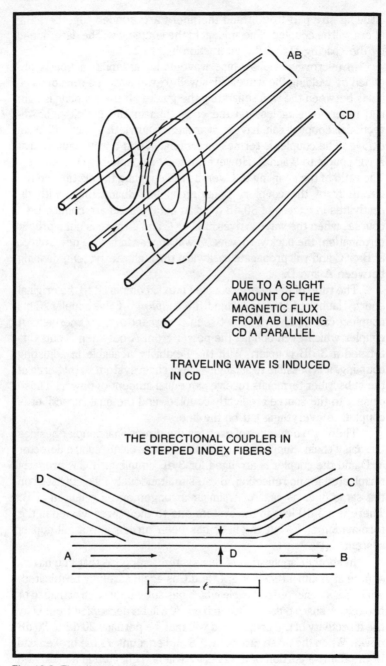

DUE TO A SLIGHT
AMOUNT OF THE
MAGNETIC FLUX
FROM AB LINKING
CD A PARALLEL

TRAVELING WAVE IS INDUCED
IN CD

THE DIRECTIONAL COUPLER IN
STEPPED INDEX FIBERS

Fig. 19-3. The directional coupler.

331

cladding stripped from it and the fibers are bonded together, the fibers will be coupled. The amount of the coupling will be determined by the spacing D and the interaction length 1.

In microwave hollow-pipe waveguide, a similar action is obtained by welding the waveguides wall-to-wall with a series of small holes between the wall separating the guides. If the coupling is light and tapers—being less on the ends than in the middle—the directional coupler can have a working bandwidth approaching an octave. The coupler is termed directional in that a wave launched at A will couple to B and C. However, it will be coupled to D much less. The ratio of the coupling to C versus the coupling to D is termed the *directivity* of the coupler. Commercial broadband units with directivities in excess of 30 dB over an octave band are common. Of course, when the wave arrives at B or C and if there is not a proper termination, the backward wave(s) will appear to have a new source at B or C and will propagate backward through the system, dividing between A and D.

The ratio of the energy coupled into C compared to the original energy launched at A is termed the *coupling* of the coupler. This coupling can be any value between 0 percent and 100 percent. A coupler which even divides the power from A between B and C is termed a 3 dB coupler. With the flexibility available in adjusting coupling, it is possible to arrange a daisy chain system so that each of the subscriber terminals receives an equal amount of power. Those closest to the source are lightly coupled, and the final subscriber is coupled to everything left on the line.

There are a number of uses for the directional coupler besides the daisy chain. Suppose that you place a source at A and a detector at D and the coupler is arranged for 3 dB coupling. If C is properly terminated for no reflection, B can simultaneously talk and listen on the same fiber as can A. A similar arrangement is required at B. There is a 3 dB loss in the system due to the power wasted in the termination at C, but the fiber has been turned into a full duplex system.

In another application of this system, suppose that you have a source at A and a receiver at D and C is again carefully terminated, while B is connected to a cable which has somehow been damaged or broken. A sharp pulse launched from A will be decoupled from D by the directivity of the coupler and will thus be perhaps 20 dB to 30 dB down. When the pulse races down B and encounters the broken end of the fiber, a portion of the energy will be reflected in a backward wave which races back toward A and D and divides evenly. The time span between the start of the pulse and the arrival of the echo at D will be directly proportional to twice the length of the system to the break, divided by the velocity of propagation. When viewed on an oscilloscope, this delay can be measured. In fact, the face of the oscilloscope can be calibrated directly in distance to the fault. This technique is referred to as *time domain reflectometry*. It is widely used in rf systems to locate cable faults and bad connectors or any other form of mismatch or reflection-producing flaw on the line. It is also extremely handy in aircraft and shipboard work for pinpointing the exact location of a fault in a cable which may be snaked through the vehicle structure.

There are certain rather significant differences between the optical fiber and the metal pipe waveguide when it comes to building directional couplers. The first and foremost has to do with the very tiny size of the fibers. The second has to do with the fact that the metal pipe waveguide is a single-mode device, whereas the optical fiber generally is not. The presence of multiple modes, perhaps numbering in the hundreds, makes the calculation of an appropriate

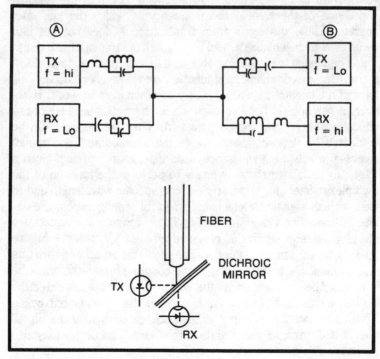

Fig. 19-4. A full duplex rf system.

coupling much more difficult, if not impossible. The dimensions that must be handled on the fiber are also at the limit of ordinary fabrication techniques. Such couplers have been fabricated by a number of experimenters; however, a commercially available line of these components is not yet available.

MULTIPLEXERS

An item that does not seem to be widely developed for fiber-optic usage is the duplexer or multiplexer. "Windows," or areas of low-loss transmission, appear in the characteristics of many of the available fibers at 0.82, 0.9, and 1.06 μm. Because communication is often a two-way street, you wonder whether it would not be possible to arrange for two-way communication simultaneously over the same fiber simply by using two frequencies. A could transmit on 0.82 and receive on 1.06, and B could transmit on 1.06 and receive on 0.82. Figure 19-4 shows such a system designed for rf operation. Each unit is coupled into the connecting line by means of a filter that contains a parallel trap tuned to the frequency at which the unit does *not* operate and a series element arranged to cancel the residual impedance of the trap at the frequency at which the unit *does* operate. Thus, the signal from transmitter A does not get into receiver A to desensitize it, and the signal from transmitter B does.

The advantages of being able to use a single fiber to perform more than one communication function are relatively obvious. This form of multicoupled system is fairly common in radio work. However, it has not yet become common in fiberoptics because of the difficulty of arranging the couplings. An interference filter can be fabricated by depositing successive thin (sometimes monatomic) layers of a reflecting substance with appropriate spacing. Such a filter can be fabricated fairly easily to perform the function of the multiplexer filter, namely to reflect one specific wavelength and to pass without significant loss another. In this configuration, the device is known as a *dichroic mirror*. These mirrors are sometimes used for the color separation required for color TV cameras. Figure 19-4 shows such an arrangement. The problems associated with this arrangement are naturally due to the coupling losses that must be introduced between the end of the fiber and the source and detector. To be effective, it would seem likely that the arrangement would have to be included within a single package containing the pigtail fiber, the dichroic mirror, and the two diodes. This device may also be forthcoming.

Another approach to this problem could use a technique

currently being investigated for solar cell construction. In certain of these devices, a complete diode is deposited upon the face of another diode. The front diode is transparent to the wavelength of the rear diode. If a Burrus diode could be fabricated in this manner, the front diode could be the LED source and the rear diode the detector.

THE TRANSITION SECTION OR SIZE ADAPTER

One of the main sources of junction losses and coupling losses between fibers and diodes stems from the size or numerical aperture mismatch. In rf work, it is common to use adapters to connect large cables to small fittings; however, in fiberoptic work you seem to be stuck with the fiber size that happens to be there. It seems likely that it should be possible to develop graded or stepped index transition sections which could greatly reduce these losses. A simple tapered transition is shown in Fig. 19-5, which adapts from a rather large lensed diode to a small fiber. Probably the principal optical difficulty to a device of this sort is that the large end will accommodate a number of modes that cannot be accommodated by the narrow end. Therefore, there would have to be a significant degree of mode conversion to take place between the wide end and the narrow end. However, note that such a mode conversion does in fact take place within a waveguide horn that is properly designed. It then seems possible to provide such transitions for fiberoptic work.

On the technical side, there is the problem of just how to fabricate such a device. The fiber is fabricated by drawing an original rod, which is quite large, down into the narrow fiber. However, the production of a sharply tapered fiber could pose some very difficult technical problems.

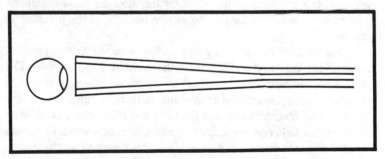

Fib. 19-5. The tapered transition section.

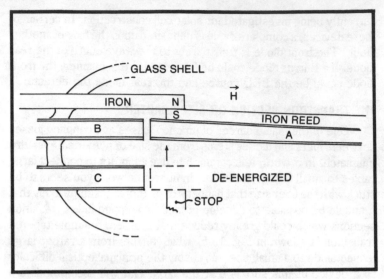

Fig. 19-6. The fiberoptic reed switch.

THE SWITCH

One of the most common items in electrical networks is the switch. For relatively obvious reasons, using fiberoptics *within* a telephone-type system would be severely restricted without an adequate switch. If every time that a switchover is to be made, you have to go from the optical to the electrical and then back to the optical level in the system, the usefulness of fiberoptics for telephone exchange or switched network applications would be nil. Ideally, the fiberoptic switch should be electronic in nature and very fast and small. With current technology, it is possible to arrange a mechanism to move a fiber from the end of one fiber to the end of another, but this form of switch is far too slow for most telephone systems and would be far too expensive because of the mechanism involved.

Figure 19-6 represents one possible solution to this requirement. This unit is a variation on the familiar reed switch. The smallness and flexibility of optical fibers make this sort of switch possible. In the presence of an axial magnetic field, the iron reed and the attached fiber bend so that fibers A and B are connected. With the magnetic field removed, fibers A and C are connected. For fibers below 125 μm in diameter, it seems likely that such a switch could be made with the glass envelope about $^3/_{16}$ inch in diameter and no more than an inch long. Allowing for the added mass of the fiber, such a device could probably be made to switch in 1 or 2 milliseconds, which would be adequate for many applications.

OPTICAL INTEGRATED CIRCUITS

A number of organizations have been investigating the construction of monolithic integrated circuit-type devices in which switching, gating, and amplification functions can be done with streams of coherent light. In many of these devices, the optical waveguide on the chip looks a great deal like a microwave stripline circuit. With the chip performing the various functions at the optical level, designers hope that tremendous increases in operational speed can be achieved, which would hold the possibility that ultrafast computers could be constructed in which the electrical supply would consist of only a power source. Also, all the logic functions would use light as a working fluid. There is no exact counterpart of the capacitive loading which tends to slow the speeds of electrical gates, so a true all-optical gate or flip-flop would probably be several orders of magnitude faster than the current electrical components.

Direct optical amplification can currently be performed with certain types of lasers that have certain characteristics similar to the microwave-traveling wave tube. However, these devices are rather large pieces of apparatus operating at high power. In principle at least, it seems possible that direct optical amplification could be performed within something like an ILD. If an integrated optical amplifier can be achieved, then at least OR gates and flip-flops should be possible. The outcome of these investigations remains to be determined, but if true optical logic does eventually materialize, fiberoptic guides will probably be the paths over which the various functions and devices will communicate.

20

Some Practical
Fiberoptic Applications

Having just discussed the basic principles of fiberoptic devices and some of the future hopes for fiberoptics, it is perhaps time to discuss a few practical applications for fiberoptic systems or fiberoptic gadgets. In this chapter, we shall treat some of the simpler and more straightforward applications which are commonly encountered.

When someone is first confronted with fiberoptics, there is usually a very strong initial enthusiasm over the vast possibilities of this new medium which eventually wears off into the question: "What can *I* use it for?" The initial applications usually use one or another of the unique advantages of the medium. A few of these will be presented in a straightforward and practical manner.

THE NONCONDUCTIVE LINK

One of the unique properties of fiberoptic cable is that it is not a conductor of electricity. Making use of this property renders a number of previously very difficult feats possible. A case in point is the measurement of radiation patterns on a broad-beamed antenna.

When someone attempts to measure the radiation patterns of an antenna which is very broad-beamed or electrically small, there is the traditional difficulty with the coaxial cable used to carry the signal to the recording apparatus.

The basic measurement of radiation patterns from an antenna is performed in a manner similar to the arrangement shown in Fig. 10-5. A remote source is set up, preferably well above the earth, with a transmitter of the desired frequency. The antenna to be measured is set up on a turntable mechanism which may rotate it on one, two, or three axes. The signal is then measured as received. In elaborate setups, a recorder may rotate a sheet of paper under the pen, according to the rotation of the turntable. The pen draws a radial plot of the received signal strength. In many of these turntables, a portion of the rotating and supporting structure is made of fiberglass and driven with silk belts in order to keep all conductive items away from the antenna.

The reason for being so particular about the conductive material is that any conductor near the antenna will cause reflections that can distort the pattern severely. If you live some distance from the local TV stations, try setting up a small TV running only on "rabbit ears." As you walk about the set, you are likely to find certain places where you can stand and make the picture noisy or perhaps even nonexistent. The reason for this is that the set is receiving not oné but several signals. One perhaps comes straight from the station, and the second is being reflected off your body. When the picture becomes torn or noisy, it means that these signals are nearly equal in amplitude and out-of-phase; therefore, they cancel one another.

To get the signal from the antenna to the recorder, it was generally necessary to drape a coaxial cable from the antenna. More than one antenna man has found that he could make the antenna pattern look like just about anything depending upon what he did with the feed cable. This does not increase any faith in the results of the measurement.

EUREKA!

On to the stage come fiberoptics. Now we have the possibility of building a nonconductive data link to get the measurements from the antenna to the recorder without any conductors to distort the radiation patterns. Furthermore, once the device is in hand, it can be used for a great many other problems, which will be discussed shortly.

In the measurement of antennas, it is common to have the source square wave modulated; that is, turned clean on and off at some audio rate, of which 1 kHz seems to be a favorite. The reason for this is that it is easier to handle the known, fixed modulation frequency and filter out the noise from the signal than it is to try to handle a continuous and unvarying signal. With a narrowband audio filter, a large portion of the noise and interference can be rejected. A similar theorem applies to the testing of optical detectors and sources. The light train can be modulated by electrical means or by some such simple technique as passing the light beam through a chopper or even the blades of an electric fan.

For antenna measurements—and for certain optical measurements—the detector of choice is a bolometer. The bolometer can be an elaborate, purchased detector or it can be as simple as a 5 mA Littelfuse® normally sold for instrument protection. This fuse is driven with an adjustable constant current source set for a current that is strong enough to make the Littelfuse heat somewhat but not burn it out. For the 5 mA Littelfuse, a current of about 4 mA seems good in most applications.

Figure 20-1 shows the circuitry for this arrangement. U1 is the. constant current generator feeding the bolometer through the pass transistor. At the temperature assumed by the tiny platinum fuse wire, the resistance of the bolometer is approximately 300 ohms. When rf is coupled into the bolometer—or for that matter, if light is shone upon the bolometer wire—the power dissipated in the wire will warm the wire and raise its resistance. Because the generator maintains a constant current through the now higher resistance, the voltage drop across the bolometer changes slightly by Ohm's law. Note that if the source were not modulated, any voltage would have to be balanced due to changes in temperature out of the measurement. The temperature change for a low input power, either rf or optical, is very small. For this reason, even mild zephyrs could alter the reading.

The wire in the bolometer is about 1.5 mm long and only 0.0001 inches in diameter (2.54 μm). It therefore will cool and warm fairly rapidly. The unit will easily follow 1 kHz square waves but starts to run out of gas at 2 kHz. We have therefore elected to modulate the source at 1 kHz. U2 constitutes an active filter with a pass band of about 10 Hz at 1 kHz. This is the item which pares most of the noise. It operates at unity gain at 1 kHz, and the gain falls to low levels very rapidly in either direction. U3 is simply an AC amplifier to bring the signal level up 100 times. The signal is then rectified by the full-wave bridge and filtered to turn it into a steady state DC with a range of 0 to 100 mV.

Up to this point, the device could as easily be used for optical detection as for rf work. The wire is tiny and it is probably necessary to use a lens to get the light on to the wire. However, a hole could be drilled in the glass case of the Littelfuse and a fiber pigtail could be epoxied in for optical measurements. The system as shown will typically have a noise floor of about −80 dBW so it is not as sensitive as a photodiode. Nevertheless, the output voltage will be directly

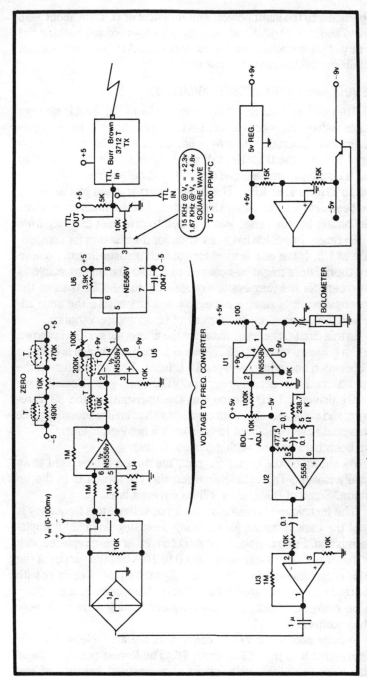

Fig. 20-1. Photo-isolated antenna pattern detector.

341

proportional to the input power, either optical or rf, from about −75 dBW to about −30 dBW, so the unit is well suited to checking the linearity of phototransistors, modulators, etc. At inputs above about −30 dBW, the bolometer burns out!

THE VOLTAGE-TO-FREQUENCY CONVERTER

The portion of the circuit in the upper tier of Fig. 20-1 is simply a voltage-to-frequency converter. U4 is a high common mode rejection amplifier operating at unity gain. U5 is a scaling amplifier arranged to scale the 0 to 100 mV input signal to the range 2.3V to 4.8V. This range was selected to give U6 an output frequency range from 15 kHz to 1.67 kHz. The V/F converter could just as easily have been arranged to swing the frequency of the timer over some other range, but the range was selected to give best linearity from the 566 timer. Provision has been made for thermistors for compensation of U5. If the unit is used for outdoor measurements, one or more thermistors might be required to render the unit as stable as 0.5 percent for the temperature range - 10° to +80°C. Without the compensation, it is usually necessary to compensate the zero adjustment for every 5 or so degrees of temperature variation.

In this circuit, the transmitter for the fiberoptic system is shown as a Burr-Brown model 3712T. This is a small unit which is built for 5V operation and which accepts TTL input. The transistor on the output of the 566 function generator/VCO could just as easily have directly driven a LED source, but the transmitter, the 3212 receiver, and a 20-foot cable with connectors had an introductory price that would have been hard to beat with a homebuilt unit.
have been hard to beat with a homebuilt unit.

As shown on the circuit diagram, the device can be used as an optically isolated DC voltmeter when the U4 switch is in the up position. Some of these uses will be covered later.

The technique of translating the input voltage to a frequency is one of the easiest means for encoding an analog voltage for digital transmission. For example, if the A/D converter were replaced with one that would operate linearly from 0 to 10 KHz, the output of the 3212R receiver could directly drive a digital frequency counter with five digits of resolution (but not accuracy). Alternatively, the voltage can be converted back from the frequency using a frequency-to-voltage converter.

A large number of V/F/V converters are available with prices ranging from less than $1 to about $6. The lowest priced of these single-chip, monolithic units offers a guaranteed linearity of one percent of full scale and a temperature coefficient of 150 PPM/°C. At the high end of the range, you can find units with a linearity of 0.01 percent and a *tempco* of 25 PPM/°C. Under the circumstances, it is scarcely attractive to try to design a unit if the accuracy is at all important with changing temperature. These units can be used for either conversion or deconversion. The temperature compensation of a homebrew unit can be very time consuming. However, for the sake of completeness, the simple F/V converter of Fig. 20-2 is offered.

The operation of the unit is relatively simple. An external clock and two flip-flops provide an output pulse of constant width. On the falling edge of the data square wave, flip-flop U1 is set through the differentiator circuit on the preset terminal. The rebound diode prevents the preset terminal from rising above V+ on the rising edge of the data pulse. At the next occurrence of a rising clock, U2 is set and U1 is cleared since its data terminal D is grounded. With the

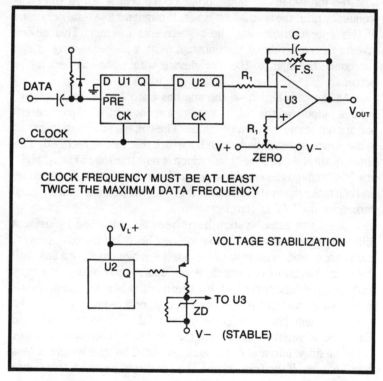

Fig. 20-2. Frequency-to-voltage conversion.

Q output of U1 low, the next rising clock pulse will set the output of U2 low. It will remain low until the next cycle occurs on the data line. The purpose of this is to provide an output pulse whose width is precisely controlled. The data clock can be controlled by a quartz crystal if necessary, and its period can be held to a few parts per million over very wide temperature ranges without any fancy adjustment. The width of the output pulse will always be precisely one full cycle of the clock in width.

The function of U3 is to serve as an integrating or averaging device. Whenever Q is high, the circuit attempts to drive a current into the inverting input of U3, which causes the output to swing and capacitor C to charge. U3 is exposed to a series of pulses whose width is determined by the clock cycle and whose repetition is controlled by the data. The output of U3 is therefore linearly related to the average voltage output of U2. After the frequency of the V/F converter has been adjusted to track the input voltage, the signal can be applied to the F/V. At the low data rate, the FS resistor is adjusted for 100 mV or some other convenient level. At the high-frequency rate, the zero adjust is set. It might take several iterations of this operation because the adjustments interact. This device inverts the rate/voltage relationship. With a proper setting, it can reproduce the input to the V/F device within one-half percent or better.

As shown in Fig. 20-2, the unit has a problem in that the "low" voltage output and the "high" voltage are not controlled parameters and are not stable with temperature. The voltage stabilization circuit in the figure shows a technique by which this can be rectified. The zener diode stabilizes the "low" which is now translated into a "high" at a fixed voltage above a stable low reference. With a little patience, you can make the system track within one-half percent with an initial zeroing of the F/V at temperature.

Could this same system have been accomplished by using a telemetry radio transmitter and receiver? Sure, but the cost, power, and space considerations would make the achievement of a link with the same freedom from interference stability and reliability a very challenging, engineering task. Furthermore, when the antenna to be measured is on a scale model of an aircraft or a re-entry vehicle, the space problem becomes quite substantial. The entire V/F and fiberoptic transmitter, along with the hearing aid batteries to power it, can be fitted into about 2.5 cubic inches. The total weight is less than 3 ounces. If the transmit LED were directly driven, this figure could even be improved.

MEASUREMENTS IN STRONG ELECTRICAL FIELDS

In the development of high-power electrical equipment, it is frequently desirable to know the temperature of certain items in the system while the system is in operation. For example, Fig. 20-3 shows a portion of an rf power amplifier of a high-power radio transmitter.

Often, the temperature of the tube envelope or other components needs to be measured *while* the system is in operation. The problem is that the tube anode may be operating at a DC potential of 35,000 volts or more and the electrical fields within the cabinet can measure in thousands of volts per meter. Even if the measuring device is isolated with a transformer or photoisolator, any wire connecting to the outside world would pick up hundreds of volts of induced field. In cases such as this, advantage is not only taken of the dielectric nature of the cable but also of the freedom from interference. There are a great many problems in high-voltage power engineering which will also yield to this approach.

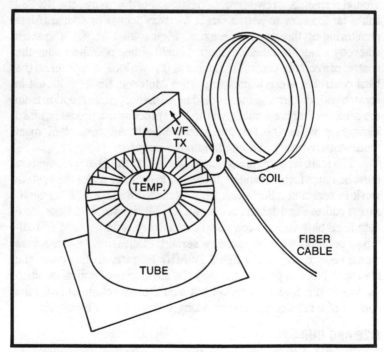

Fig. 20-3. The fiberoptic link can be used for measurements in areas where high voltages and high electric fields are present.

For example, one of the most difficult measurements in a high-voltage power line system is the onset of corona. A simple current probe imbedded in the insulator structure that supports the cable can make the measurement, but the entire instrument is then operating at perhaps 350 kV, and there is no way to get the reading to the operators. Again, the fiberoptic datalink comes to the rescue. However, when a fiberlink is to be used in a measurement of this sort, a series of high pot and tracking tests of the fiber to be used is absolutely necessary.

COMPUTER DATA LINKAGE

With the growing popularity of microprocessor control, there is a growing requirement for the linkage of a remote, processor-controlled machine with a central data terminal, etc. The form of link has been realized in the past, mainly via RS-232 standard links. However, in something like a steel mill, where there are a large number of electric arc furnaces, it is not unusual to find that there may be potential differences as high as 35 or 40 VAC between the ground terminals on two remote devices. Furthermore, the RS-232 link is far too slow to permit direct memory access or address/data monitoring at the remote terminal. Figure 20-4 shows a system whereby a remote machine control could either operate under the control of a remote central computer or its own local computer. If the local controller were something like a Motorola 6802 with built-in scratch pad, the arrangement could be such that an interruption from the remote machine would give the local processor a predetermined idling loop while the remote machine assumed control of main read/write memory and input/output functions.

The main facet of this arrangement is that the fiberoptic system must be much faster than the processor in order to make the system work in real time. For example, a standard 6802 will, in general, assert address and data lines every 2 μs. There are 16 address and 8 data lines plus various controls such as read/write, etc. For real-time operation without pause, a serial transmission of these lines would require something like a 32-MHz bit rate. In this case, the fiberoptic link can provide not only the advantage of optical isolation for freedom from interference, but also the advantage of real-time control of a remote processor, along with real-time monitoring.

SIZE AND DURABILITY

One of the advantages of fiberoptic cables is the extremely small size of the fibers. Often, there is a requirement to monitor something in a location which is too tiny to insert a practical light source or light sensor. One of the most commonly encountered applications of fiberoptics to these problems is found in your dentist's office. A number of dental drills have fiberoptic cables that conduct light from an external source down to the point of the drill where the dentist obviously needs it.

An application of this type is shown in Fig. 20-5, in which a sparkplug has been drilled to accept an optical fiber. The fiber is used to conduct the combustion flash within the cylinder to the fast spectrophotometer. This spectrophotometer is basically similar to the unit shown in Fig. 11-3, except that instead of mechanically using the motion of the mirror to direct the light to a fixed slit and a single detector, the detector consists of an array of photodiodes. The fiber itself serves as the input slit and no output slit is required since the photodiode array consists of 211 photodiodes and a scanning mechanism. The optics and electronics for this unit have been developed for the Hewlett-Packard 8540A spectrophotometer. Photo arrays for the ultraviolet 200 to 400 ηm and the visible 400 to 800 ηm ranges

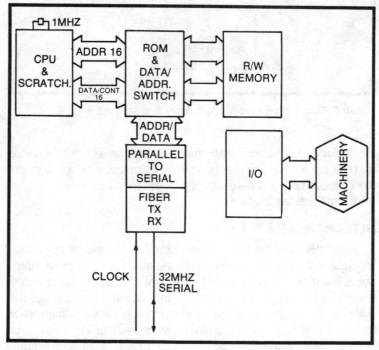

Fig. 20-4. Remote microprocessor monitoring and control.

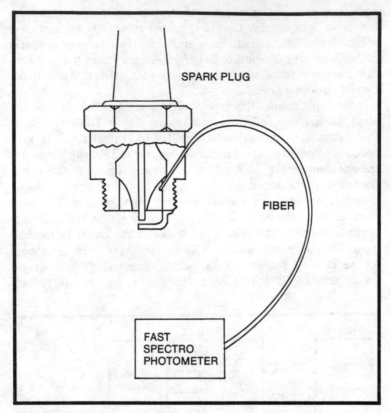

Fig. 20-5. Combustion spectrophotometry in a gasoline engine.

are included within this instrument. Observations within these spectral ranges yield an insight into the combustion process within the cylinder under various conditions of load, pressure, temperature, and fuel/air mixture.

MECHANICAL ISOLATION

One very difficult type of measurement to make is the measurement of angular vibration on rotating machinery. Angular vibration is often caused by changes in the point of contact of gear teeth, eccentricity of the gears, and other mechanisms which make the angle through which a shaft turns in a period of time anything other than a straight line curve. The main problem with this measurement is that the measuring device—a strain gauge or an accelerometer—is perhaps turning quite rapidly. The measurement can be very important since angular vibration can easily build up by resonance so that the operation of the machine is rough and shaft and part fatigue and breakage frequently result.

For measurements of this sort, advantage can be taken of the fact that the optical fibers do not have to be in direct mechanical contact to be well coupled. Figure 20-6 shows an optical rotary joint which will permit this type of measurement. The battery-powered sensor and transmitter box are mounted on the rotating machine, and the optical fiber is brought down to the center of the wheel. The optical rotary joint consists simply of a light shield which loosely encases the transmitter fiber and a second stationary fiber which connects to the receiving apparatus.

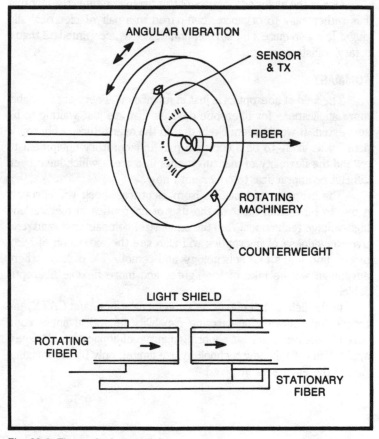

Fig. 20-6. The optical rotary joint.

This sort of thing can also be done with slip rings making electrical contact. But it is very difficult to construct a set of slip rings which is sufficiently free of noise or erratic contact. By contrast, it is relatively easy to obtain a good, smooth optical rotary connection. If the fiber fits too loosely in the light shield, there may be some modulation of the light beam with rotation. This is not a serious problem, however, if the link is digital. If an analog link is to be used, it will generally be found that sufficient freedom from modulation can be obtained if the gap between the fibers is left relatively large. This increases the insertion loss but permits the beam to broaden out from the transmitter fiber so that the receiver fiber is riding on the broad nose of the radiation pattern. It is relatively easy to hold the "wow" of the rotary joint to less than 0.1 dB on a joint turning at 3600 rpm.

One of the major advantages of this arrangement is the fact that it is rather easy to organize, compared to a pair of electrical slip rings. It is also quite a bit smaller and easier to accommodate than a rotary transformer.

SUMMARY

The field of fiberoptics is just in its infancy. There are probably more applications for fiberoptic devices that are just waiting to be suggested. It would seem likely that in the near future, a fiberoptic data link is likely to be a common part of laboratory equipment to extend the flexibility of measurements into areas which have been difficult or impractical to cover until now.

The property of isolation from electrical shock will certainly prove to be a chief facet of the use of fiberoptics in medical and high-voltage techniques. And the small size, compactness, and relative ruggedness of fiberoptics will also see the extension of techniques into areas like seismology and remote TV pickup, where advantage will be take of the lighter and more flexible fiberoptic cable.

In the field of telephony, telecommunications, and CATV, the tremendous information-carrying capability of optical fibers compared to their size will surely bring a surge of application. The new applications of this new technology are limited only by the imagination of the engineers involved.

21

A Low-cost Transmitter and Receiver Project

One of the major differences between a textbook and a handbook is that the handbook contains some practical projects. In an electronics handbook, this usually means that a number of circuits are included with practical values in the circuits so that they can be built by the experimenter. In previous chapters, a number of such circuits were presented. This chapter serves to introduce a low-cost fiberoptic transmitter and receiver.

The link which can be constructed using these components is relatively slow. However, the bandwidth is more than adequate for audio analog transmission and will handle digital transmission to rates on the order of 50 kHz. This rig is useful for a number of purposes, besides the direct primary use of fiberoptic transmission. For example, it can make NA and acceptance cone measurements. It is also useful in making a variety of tests on optical fibers, as will be described later. In addition to this, it can frequently be used in IR experiments and measurements which do not involve fibers at all but rather depend upon transmission through air.

THE TRANSMITTER

The transmitter is shown in Fig. 21-1. This is a relatively simple affair involving only a single transistor. It is essentially similar to the transmitter shown in Fig. 16-1, except that it uses a PNP transistor. The LED is the Motorola MLED 900, which is a low-cost plastic-packaged IR LED. The package proper is a small crystal-clear affair with the lens molded in and radial leads. The LED has a symmetrical

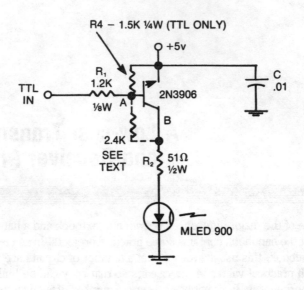

R4 — 1.5K ¼W (TTL ONLY)

+5v

R_1
1.2K
⅛W

TTL
IN

A

2N3906

B

C
.01

2.4K
SEE
TEXT

R_2 51Ω
½W

MLED 900

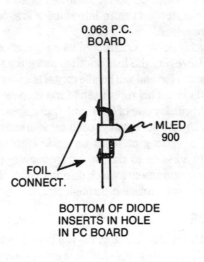

0.063 P.C.
BOARD

MLED
900

FOIL
CONNECT.

BOTTOM OF DIODE
INSERTS IN HOLE
IN PC BOARD

Fig. 21-1. The project transmitter schematic, construction details, and PC board foil layout.

FITTING HOLDER
- EPOXY TO P.C. BOARD
MATL - PVC ROD.

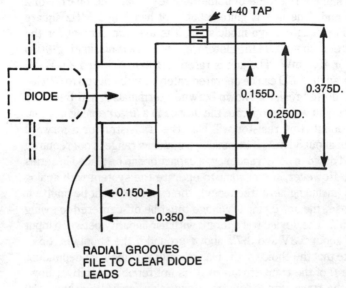

4-40 TAP

DIODE

0.155D.

0.250D.

0.375D.

←0.150→

←——0.350——→

RADIAL GROOVES
FILE TO CLEAR DIODE
LEADS

PC BOARD
FOIL LAYOUT

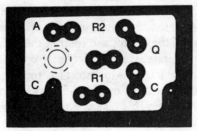

spectrum centered at 0.9 μm with half-power points at 0.86 and 0.94 μm. The radiation is quite invisible to the human eye. You can't tell whether the diode is operating or not in a dark room.

With a typical TTL drive, the transistor will turn the diode completely off and completely on. With square wave TTL input, the unit will show a rise time of something less than one-quarter of a microsecond. The unit is quite capable of handling 1-MHz square waves. For square wave modulation, the average current for the transmitter is about 0.038A; therefore, the diode is dissipating 0.038 $\times$ 1.2V or 45.6 mW. The unit is rated at 120 mW with a 2 mW/°C derating, so the LED can be operated safely in ambients up to 62°C.

The dotted resistor shown between terminals A and B is used *only* when you want to operate the device in a *linear* analog fashion. With no input, this resistor will bias the transistor for a forward current of about 38 mA. If the analog frequency range is not required to extend down to zero frequency, a capacitor can be placed in series with R1. However, if you want to operate the system with analog voltages (including zero frequency), the capacitor must be omitted. In this case, the analog input voltage must be off-centered to swing about 4.5V. The device will operate with fair linearity between input limits of about 5.3V and 3.7V above ground if the supply is +5V.

Note that the choice of the bias resistor is somewhat dependent upon the β of the transistor used. It is not terribly sensitive; however, if the transistor β departs appreciably from the value of 50 measured in the model, then a slight revision in this resistor is a must. To do this, simply adjust the value to take the transistor current to 38 mA. A smaller resistance value will raise the current, and a larger value will decrease the current.

If DC analog operation is required, it is important that the supply voltage be closely regulated. A change in the supply voltage will shift the quiescent point of the circuit and restrict the range of swing that the transistor can tolerate. It is usually best to operate the unit for analog modulation so that it begins to distort by "topping out" at the same time that it "flat bottoms." For a high-fidelity reproduction, a point not far removed from this bias will tend to minimize distortion. A THD of less than 0.1 percent should be achievable with the transmitter on analog modulation with proper centering.

Because the unit is made to operate at relatively high frequencies, the PC board layout should be kept as compact as possible. The layout shown on the figure provides the rise times described. One feature of the board is the fact that a 0.150-inch hole is drilled through the board to permit the diode to sit down flat against the leads.

For a general-purpose instrument transmitter of this sort, it is somewhat difficult to arrive at a holder which is suitable for a wide variety of purposes. The design shown in Fig. 21-1 was evolved with flexibility in mind. If the only application of the unit is to be in a fiberoptic link, then the holder can be modified to take the selected fiber directly as shown in Fig. 16-1. Alternatively, fiber holding bushings can be turned to fit any particular fiber or fiber bundle. This arrangement allows the unit to be used for testing fibers. Different bushings can also be turned to accommodate a variety of connectors as well. Thus, it is possible to accommodate assembled cables with fittings attached.

The fitting shown is turned out of polyvinyl cloride (PVC) rod since this is an easily turned opaque plastic. A radial slot to clear the LED leads is required. Because the PVC is a good insulator for electricity, it is not necessary to otherwise insulate the base of the fitting. The fitting can be glued directly to the PC board, but be careful not to get epoxy on the LED lens. Unfortunately, the PVC is not a good conductor of heat, so some of the usable rating of the LED is sacrificed by this arrangement. If the fitting is turned of aluminum instead, the effective temperature and power rating of the diode rises, but care must be taken to insulate the leads to avoid shorting the diode. The presence of the fitting is very handy in most experiments since it is otherwise rather difficult to obtain precise alignment of the diode.

THE RECEIVER

The experimenter's receiver is shown in Fig. 21-2. This receiver uses a silicon phototransistor for simplicity. The unit selected is the Motorola MRD 3051, which was chosen because it is inexpensive and reasonably sensitive. Motorola offers this unit in a series of grades numbered MRD 3050 through 3056. The difference between these units lies in the collector-emitter guaranteed radiation sensitivity, which ranges from a minimum of 0.02 for the 3050 to 0.4 mA/mW-cm² for the 3056.

The spectral response for this phototransistor is not specified in the data sheet, but it is probably rather close to that shown in Fig. 14-6, extending well through the visible range and peaking near 0.85 μm. The unit is not particularly fast. Motorola states the rise time as 1 μs and the fall time as 10 μs, but it notes that the rise time is measured with a 1 μs argon flash. Using square waves and the transmitter described earlier, the rise time was measured at 10 μs and the fall time as 12 μs. This would give the unit a −3 dB point in

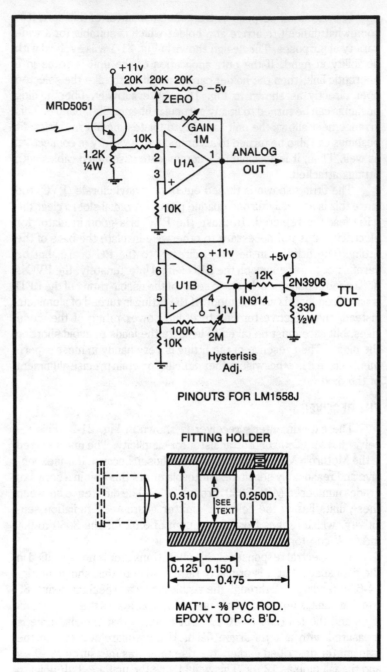

Fig. 21-2. The project receiver schematic and construction details.

the vicinity of 30 kHz. The sensitivity does not slope off too rapidly and a square wave train at 100 kHz comes through as a sawtooth wave of perhaps −6 dB sensitivity. With the lenses of the transmitter and receiver separated by 4 cm and the square wave TTL drive providing 38 mA of transmitter current, the square wave amplitude on the emitter resistor of the phototransistor measures 250 mV peak-to-peak. If a dark current of about 20 nA is considered the worst case, the power ratio is:

$$\frac{0.250v}{20 \times 10^{-9}A. \times 1200\Omega} = 10,416 \quad ; \quad \text{exists.}$$

Then dynamic range is:

$$P_{min} = 10 \text{ lag } 10,416 = -40.18db$$

And the maximum workable range is:

$$\frac{Rmax}{.04m} = \sqrt{10,416} = 4.08m$$

Remember that the power ratio is proportional to the current ratio in the phototransistor.

This sensitivity is attainable without any bandwidth filtering. If the bandwidth were restricted, then the usable range through air would rise as well.

Because this is intended to be a simple device, no great pains were taken with the design of the amplifier. For analog output, only the op amp circuit, U1A, would suffice and a simple op amp such as the 741 could be substituted. Even here, however, the inclusion of a zeroing adjustment and a gain adjustment is desirable so that the output can be centered for subsequent equipment. For TTL usage, it is valuable to include comparator U1B, which has an adjustable hysteresis. With the hysteresis adjustment at zero resistance, the hysteresis is ± 1V. With the hysteresis adjustment at maximum resistance, the hysteresis is ± 0.048V. The transistor is used to convert the output to TTL level, and the unit will drive two standard or about 10 low-power TTL devices.

The op amps were selected mainly for their low cost, availability, and ease of use. The unit is known as a 1458 or 1558 (the difference is a wider temperature rating for the 1458), or a 558 or 5558 from other manufacturers. The unit consists essentially of two 741s in a single package which can be obtained as either a TO-5 or an eight-pin DIP. Because the phototransistor is not blindingly fast, these op amps are usable and do not significantly degrade the performance of the unit overall. If a faster photodiode were selected then a more sophisticated receiver, such as the circuit of Fig. 17-5, which uses faster op amps, would be desirable. Because of the relatively slow speed, this unit can be built on a perfboard and great care is not required to minimize lead inductance.

USES OF THE UNIT

It was noted earlier that the unit can be used as a digital or analog fiberoptic link. In particular, the voltage-to-frequency converter shown in Fig. 20-1 and the frequency-to-voltage converter of Fig. 20-2 can be applied to this unit to produce a fiberoptic or through-the-air data link. With good quality fibers, of course, the range can be extended greatly over the range in air. The transmitter and receiver in this case simply assume the function of the commercial units shown in those earlier descriptions or any of the other applications discussed in Chapter 20.

In addition to this application, the use of the units for the numerical aperture measurement was discussed earlier. However, there are a number of other applications for these units. For example, the transmitter and receiver can be used as a long-range photo interruptor and all of the various photo-interruptor applications pursued.

For example, the unit can measure the rotational speed of drills and shafts on machine tools. This can be done by having the transmitter shine on the rotating element, which has been marked with a piece of black crepe tape. The TTL output can then be fed to a digital frequency counter or a calibrated frequency-to-voltage converter. During most of the rotation, the light from the transmitter bounces back to the receiver. When the tape comes along, however, the bounce is absorbed, thus giving a pulse in the output. In an alternative method, attach a tape "flag" to the rotating element. The light path is arranged so that it is interrupted only by the flag.

Used alone, the receiver is fast enough to trigger slave strobe lamps for photography. When the receiver detects the primary flash, it triggers the slave strobe for shadow filling and such.

The unit can also be used to measure the speed of a fan or a model airplane engine by having the blades interrupt the path. Also, the transmitter can be installed in a toy gun and the receiver arranged to flash a light when "hit." There are many game-type uses for the unit.

Index